Of Flies, Mice, and Men

François Jacob

Translated by *Giselle Weiss*

OF FLIES,
MICE,
AND MEN

Harvard University Press

Cambridge, Massachusetts, and London, England · 1998

Copyright © 1998 by the President and Fellows of Harvard College
All rights reserved
Printed in the United States of America
Originally published as *La Souris, la mouche et l'homme,* © Editions
Odile Jacob, 1997

Published with the help of the French Ministry of Culture

Designed by Marianne Perlak

Library of Congress Cataloging-in-Publication Data

Jacob, François, 1920–
    [La souris, la mouche et l'homme. English]
    Of flies, mice, and men / François Jacob ; translated by Giselle
Weiss.
        p.   cm.
    Includes index.
    ISBN 0-674-63111-0 (alk. paper)
    1. Molecular biology.   I. Title.
    QH506.J3313 1998
    572.8'01--DC21
    98-7289

Contents

....

Of Flies, Mice, and Men

Introduction

....

"But then to what end," asked Candide, "was the world formed?"
"To make us mad," said Martin.

—VOLTAIRE, *CANDIDE*

In the short story "The Creation," Dino Buzzati describes how the Almighty achieved his work. First he made the universe, with, in one corner, a small sphere—a planet—devised in a way that permitted a very curious and amusing phenomenon: life. "The notion of this little sphere, suspended in the immensity of space and bearing a multitude of beings that were coming into existence, growing, multiplying, and dying, struck the Creator's fancy."

A host of architect-angels immediately rushed to submit models for the innumerable species of living things—plants and animals—needed to ensure the success of the planet. After many discussions with his executive committee, the Almighty ended up approving most of the projects. There remained, however, one unfortunate angel who had failed to attract the attention of the Lord. Finally he managed to thread his way to the feet of the Creator. His sketches showed an animal whose appearance was truly disagreeable, not to say repugnant, but arresting because it was so totally different from the other designs. On one side he drew the male; on the other, the female. His animals had four legs like other animals, but—at least to judge from the drawings—they only used two legs for walking. No fur except for a few tufts here and there, mainly on their heads. The Creator was not very enthusiastic. The designer insisted: it would be an exceptional invention; it would be the only rational being, the only one capable of worshiping the Creator, of erecting temples in his honor, of waging dreadfully murderous wars in his name. "You mean to say an intellectual," responded the Almighty in horror. "Anything but that!" The inventor of man and woman went away grudgingly.

The earth took shape, a place of marvels and cruelties, ecstasies and fears, love and death. It became filled with a variety of living things, pleasing and odious, gentle and savage, terrible and mundane, beautiful and repugnant: the centipede, the oak, the tapeworm, the eagle, the mongoose, the gazelle, the rhododendron. The lion! Night fell. Tired but content, the Lord dozed off. Suddenly he felt someone tugging at the sleeve of his mantle: it was the tiresome designer of the human couple returning to the attack. What a crazy idea, even dangerous! thought the Creator. Yet what a fascinating game, what a temptation! And, half asleep, he accepted the fateful project.[1]

It requires a lot of ingenuity—you might even say perversity—a lot of knowledge acquired in the face of all perceptible evidence, against all intuition, to be able to discern common features among the extreme variety of life forms. At first glance, the animals that came off

1. D. Buzzati, D. "La Creation," in *Le K.* (Paris: Robert Laffont, 1967).

Noah's Ark appear to have been conceived individually by an Artist who found creative expression in assigning features to the diversity of living beings. Stranger still are certain attributes—indeed we can scarcely imagine what motivated the Artist to make them the common fate of living beings—such as being born, growing up, and aging. In particular, the necessity of populations to disappear little by little by dying and to renew themselves by reproducing.

From its inception at the beginning of the nineteenth century, biology was preoccupied with structure and function. Despite the clamor of those who argued for the indivisibility of living beings, reductionism achieved victory after victory. Yet the deeper biology probed differences among organisms, the more they faded away, affirming a unity of living things. In the middle of the last century, study of the cell revealed the structural unity, in a sense, the atom, of life. Next came the theory of evolution, and with it a common origin. Before the Second World War, biochemists demonstrated commonalities of structure and function underlying the diversity of life forms. From the 1960s onward, molecular biology provided evidence for the unity of genetic systems and the basic mechanisms that regulate cell function. And finally, with the advent of genetic engineering in the 1970s, the unity of the living world was proven to a point not previously imaginable. All the creatures that inhabit this earth, whatever their environment, size, or means of subsistence—snail, lobster, fly, or giraffe—all turn out to be made from molecules that are more or less identical. And likewise, from yeast to humans, there are groups of closely related molecules that serve to assure universal life functions, such as cell division or signaling between the cell membrane and the nucleus.

So now a fantastic paradox emerges: organisms of the most different sorts are constructed from the very same battery of genes. The diversity of life forms results from small changes in the regulatory systems that govern the expression of these genes. The structure of the adult organism is determined by the development of the embryo. If, during that development, a gene is expressed a little earlier or a little later, or if it functions more abundantly in slightly different tissues, the final product, the adult animal, will be profoundly modified. Which is how it happens that, despite their enormous differences, fish and mammals have approximately the same genes, just as do crocodiles and sparrows. The creative potential of the regulatory networks is owed to their hierarchical and combinatorial nature. Quite a considerable variation of animal forms can be produced at several levels simply by tinkering with the network of the numerous regulatory genes that determine the moment at which one gene or another is expressed. It is the similarity of genes governing the embryonic development of very different organisms that ultimately made the evolution of complex life forms possible. If each new species had required the creation of new regulatory networks to make its appearance, the time required for evolution, based on paleontology, would have been too great. Evolutionary tinkering is what allowed regulatory elements to combine into highly varied developmental systems.

It appears, then, that all life forms are constructed with the same modules, distributed in different ways. The living world is a sort of combinatorial system composed of a finite number of parts, like the product of a gigantic Erector set; it is the result of a ceaseless process of evolutionary tinkering. This radical change of perspective in the field of biology occurred only in the last few years.

It is hard for most people to understand how researchers can be interested in what seem to be totally uninteresting questions. Why don't scientists concern themselves with truly important problems, such as the nature of life and death, the common cold, or hair loss? But no researcher has ever managed to become famous by finding nothing. In the words of Peter Medawar, "if politics is the art of the possible, research is surely the art of the soluble."[2] In fact, scientists apply themselves to what they believe to be the most important of the problems that seem tractable, that is, those that rightly or wrongly they think they will be able to solve. For their job is not just to discuss questions. It is also to find solutions to them. As in many other human activities, and in life itself, the scientist navigates between two poles, the desirable and the possible. Without the possible, the desirable is little more than a dream. Without the desirable, the possible is boring. It is often difficult to resist dreaming and utopianism. But experimentation helps to contain the imagination. At every step, scientists must lay themselves open to criticism and experimental findings in order to limit the influence of dreaming in the representations they construct of the world. Science is a process of perpetually reconciling what might be with what is.

To tackle an important problem, to have a reasonable chance of finding a solution to it, biologists must choose a suitable organism, an organism that permits them to carry out certain types of experiments required for the proposed research. At the beginning of the century, when Thomas Hunt Morgan wished to analyze heredity, he used the fruit fly *Drosophila,* which enabled him to resolve questions about transmission of traits. In the middle of the century, attention focused on the chemical nature of heredity, of analyzing the basic functions of the cell. For this, molecular biologists had to turn to bacteria, which are uniquely suited to such studies. Later, when genetic engineering allowed access to the genetic material of any arbitrary organism, *Drosophila* came back into favor. It offered the possibility of studying for the first time the genetic basis of embryonic development and the major functions of the organism. After which, the astonishing discovery of the persistence of the same regulatory mechanisms over the whole range of evolution made it possible to study mammals, in this case, the

2. P. B. Medawar, *The Art of the Soluble* (London: Methuen, 1967), p. 87.

mouse. Hence biologists who, like me, have lived through this period, have had to reorient themselves several times over.

At the end of the 1960s, it was clear that the center of gravity in biology was shifting. Although the study of bacteria and viruses still had much to teach us, it was slipping to second place. If we didn't want to stand around rehashing the same questions, we needed the courage to abandon old lines of research and old models, to turn to new problems and study them with more suitable organisms.

The word "courage" is not too strong. The daily interaction over years with a living organism, however humble, entails a certain familiarity. You could almost say that you acquire a certain tenderness for it. After fifteen years of working with a particular colon bacillus, I had accumulated hundreds of mutants. In each of these mutants, one or another of the cellular functions, many of them indispensable to the life and reproduction of the bacteria, had been altered. To abandon this work and all that it offered; to renounce the kind of intimacy that comes with the knowledge of little unwritten quirks, the folklore that surrounds the work on any one organism; to start again from zero with another, unknown organism whose idiosyncrasies I would have to discover—all this was a considerable sacrifice. It was a little like leaving a loved one. But, at the same time, the new project was an exciting one. It would mean entering an unknown world, beginning a new life, becoming young again. . . .

This book is about molecules, reproduction, and evolutionary tinkering. It is also about the way biologists work, and how they contemplate beauty and truth, good and evil. In recent years I have given a number of lectures, both in France and abroad, during which I have touched on these themes. These experiences provided me with the material that was the wellspring for writing this book.

1 The Importance of the Unpredictable

···

In ancient Greece, no king would dare risk the slightest undertaking without first consulting an oracle. But even in Greek times, ordinary mortals did not have the ability to predict the future, a power reserved for the gods, and only certain gods at that. Rarely, as thanks for services rendered, a god might give the power of divination to a mortal. Such was the gift of the most famous of soothsayers, Tiresias, who distinguished himself by foretelling the destinies of celebrities like Narcissus, Oedipus, and Hercules. Tiresias was blind. Some say that Athena blinded him because he inadvertently saw her naked in the bath. But then, moved by the wailing of Tiresias's mother, Athena drew the serpent Ericthonius from her aegis and commanded him to cleanse Tiresias's ears with his tongue so he could understand the language of prophesying birds.

Others say that Tiresias received his gift of prophecy from Zeus for having resolved one of the most difficult questions confronting humans, namely, who gets more pleasure out of making love, men or women? While walking one day on Mount Cylene, Tiresias came upon two snakes mating. The snakes were disturbed and attacked him, whereupon Tiresias defended himself with his walking stick and killed the female. Immediately transformed into a woman by Athena, he/she became a celebrated prostitute. A few years later, Tiresias chanced to be in the same place, in the same situation. Again a pair of snakes attacked her. Only this time it was the male snake Tiresias killed with the walking stick, thereby regaining his male form. A few days later, a domestic dispute broke out between Zeus and Hera, who was reproaching her husband for his countless infidelities. Defending himself, Zeus retorted that, in any case, when Hera shared his bed, it was she who had the better time. "Everybody knows," he grumbled, "that sex is much more pleasurable for women than for men." "Not at all!" cried Hera, furious. "You know very well that it's the exact opposite!" Tiresias was asked to settle the quarrel, giving his opinion on the matter based on his personal experience. "If you score sexual pleasure on a scale of ten," he said, "then women get three times three, and men only one." Exasperated by Zeus's triumphant smirk, Hera blinded Tiresias. To compensate him, Zeus granted him the gift of prophecy and a life that would unfold over seven generations.

Divination and prophecy thus provided a means of communication between men and the gods, though the gift always remained the prerogative of the gods. Things have hardly changed since then. In fact, as a consequence of a few oversights, the gods must have taken a closer look and decided to stop delegating the right to predict the future altogether. Indeed, no soothsayer, seer, or psychic foresaw the economic growth of Japan, or the fall of the Berlin wall, or the breakdown of the socialist republics, or the AIDS epidemic, or any of the other significant events that have transpired over the past few years.

Granted: we cannot know the thing that concerns us more than anything else in the world, that is, what will happen tomorrow. Yet to look ahead is one of our most common and important activities.

Paul Valéry called it "making the future."[1] An organism is alive only while it continues expecting to live, even if only for another instant. There is not a single movement, not a single gesture that doesn't imply an afterward, a later, a transition to the next moment. To breathe, to eat, to walk, is to anticipate. Seeing is foreseeing. Every one of our actions, every one of our thoughts involves us in what will be. For every human being the future merges with the very act of living.

Our imagination unceasingly projects before us an ever-changing image of what might happen, what is possible. And we are forever checking our hopes and fears against this image, altering our desires and aversions to fit it. Our organism is a sort of machine for predicting the future—an automatic forecasting apparatus. But though it is in our nature to look constantly toward the future, the system is arranged in such a way that our forecasts must remain uncertain. We cannot think about ourselves without thinking about the next instant. Yet we cannot know what that instant will bring. What we predict today will not come true. Changes will certainly occur, but the future will be different from what we had envisaged. Everything is so arranged that we cannot know the future, though that is precisely what interests us the most. This need to imagine the future and the impossibility of knowing it are woven into the very fabric of life. They form a basic element, a fundamental constituent of life. "How could we live without the unknown before us?" asks René Char.[2]

The only event we can predict with any certainty is our own death. But if the idea of death is almost tolerable, it is because, with certain

1. Paul Valéry, "Graduation Ceremonies at the Collège de Sète, July 15, 1935," in *The Collected Works of Paul Valéry*, vol. 15, *Moi*, ed. Jackson Mathews, trans. Marthiel Mathews and Jackson Mathews (Princeton: Princeton University Press, 1975), p. 274.

2. René Char, "Le Poème pulverisé," in *Oeuvres complètes* (Paris: Gallimard, 1983), p. 247.

exceptions, its hour remains totally unpredictable. Unpredictability, here, takes the place of immortality. Jonathan Swift[3] had his own recipe for making the idea of death bearable. On his trip to the island of Luggnagg, Gulliver meets a polite and generous people who have a distinctive feature: a small minority of the population, the Struldbruggs, are immortal. Immortality is not the privilege of certain families, but the result of luck. You might think that a country where every child is born with a chance to live forever would be a happy place. But you'd be wrong. Because the Struldbruggs do not acquire a state of perpetual youth with immortality—they do grow old. And just like the lover of the goddess Aurora, upon whom Zeus conferred immortality but not eternal youth, as they age, the Struldbruggs become disgusting, insufferable, and unhappy. The rest of the people, that is, the vast majority, hate and despise them. For the Struldbruggs, death would be a release from what they see as the terrible scourge of immortality.

Why do things happen one way and not some other? Why are they not predictable even if, afterward, they are explainable? This is one of the themes that Tolstoy develops toward the end of *War and Peace*. For Tolstoy, there is no correlation between events and human will, between the succession of battles in the course of the Russian campaign and the efforts of the opposing leaders, Napoleon and Kutuzov. The best example of this junction was the battle of Borodino. Neither the French nor the Russians had any reason to start it. The result was that the Russians moved a step closer to losing Moscow, which was their greatest fear, and the French a step closer to losing their army, which was *their* greatest fear. Yet well before the fight there was no doubt that this would be the outcome. That did not keep Napoleon from proposing the battle nor Kutuzov from accepting it.

3. Jonathan Swift, *Gulliver's Travels,* ed. Peter Dixon and John Chalker (Harmondsworth: Penguin, 1967), pp. 251–260.

Kutuzov had long been trying to anticipate what actions Napoleon might take. He had imagined the French army concentrated at one location to break through the Russian lines, or deployed in several companies to surround his Russian troops. He had conjectured, thought, imagined. But the one thing he could not anticipate was precisely what did happen: the crazy, fitful comings and goings of Napoleon's army. "No one at the time foresaw," says Tolstoy, "(what now seems so obvious) that this was the only way an army of eight hundred thousand men—the best army in the world and led by the best general—could be destroyed in conflict with a raw army of half its numerical strength, led by inexperienced commanders as the Russian army was."[4] Not only did no one predict the outcome, but the Russians seemed to take a perverse pleasure in doing their utmost to sabotage any measures to save Russia, while the French, despite the experience and purported military genius of Napoleon, were hell-bent on reaching Moscow at the end of summer, in other words, to do the very thing that would ruin them.

Thus, Tolstoy continues, in war leaders believe they can foresee, decide, command, and determine the course of things. In reality, everything depends on the fortuitous action of a subordinate, or perhaps on some spontaneous movement that traps or unexpectedly brings down an entire army. "The general," says Tolstoy of Kutuzov, "is always in the midst of a series of shifting events and so he can never at any point deliberate on the whole import of what is going on. Imperceptibly, moment by moment, an event takes shape" (p. 979). The best general is thus one who pays no attention during strategy sessions and who reads detective novels on the eve of battle. The wisest general doesn't act but lets events happen. "Why then did things happen thus

---

4. Leo Tolstoy, *War and Peace,* trans. Rosemary Edmunds (London: Penguin, 1957), p. 811.

and not otherwise?" concludes Tolstoy. "Because they did so happen" (p. 1342).

It is not uninteresting to note that, as this century draws to a close, the "future" business is flourishing in Western society. In France, some forty thousand seers, soothsayers, psychics, and other fortune tellers collect, annually, a sum of several tens of billions of francs. Nearly one in ten Frenchmen, including many of our political leaders, seems to be behaving like the kings of the *Iliad* by consulting his favorite psychic more or less regularly. A recent president of the United States was rumored to make decisions only with the advice of his wife, who herself first consulted her astrologist. And a few French presidents, too, have preferred to keep the stars on their side.

At the same time, so strong is public confidence in the predictive capability of science, that seers and psychics do not hesitate to describe their predictions as "scientific," the better to lure clients. And seers and psychics have no monopoly on this confidence. Every year symposia are organized on futurist themes of the genre "What will biology be like in twenty years?" or "Medicine in the twenty-first century" or "What will the effects of science on society be at the beginning of the next century?" People who choose these subjects enjoy the call of the open sea. They want to open up new routes, offer unlimited prospects into which a tractable future will fit quite naturally. Indeed, politicians and administrators of science cannot bear the idea of blind research that gropes along with no guarantee of results. Since they are supposed to be directing research, at the very least they have a duty to steer it toward some destination. They long to participate in this great human adventure. And for them, making plans, sketching out a direction, and speaking of the future is tantamount to mastering it.

Yet science is also unpredictable. Research is an endless process; we can never say how it will evolve. Unpredictability is part of the very nature of the scientific enterprise. If what we are going to find out is truly new, then it is by definition something we cannot know in advance. There is no way to say where a given field of research will lead. This is why we cannot choose certain aspects of science and reject others. As Lewis Thomas points out,[5] either you have science or you don't. And if you have it, you can't take only what you like. You also have to accept those aspects of it that are surprising and disturbing.

So it is futile to hope to predict the direction that science will take. At any instant, given what we know, we can imagine what will happen in, say, five years. But this is the least interesting part of research, the humdrum, the routine. The really interesting part is what we cannot foresee. It's what someone nobody has ever heard of, working in a cellar or attic, will suddenly prove or one day tackle from a new angle, thus throwing new light on the universe or a tiny piece of the universe. You could even say that in basic research, if a good dose of uncertainty doesn't accompany the results of an experiment to start with, there's almost no chance that the work in question is an important one. Most often we begin with data that are somewhat ambiguous or incomplete. The problem is to find relationships between seemingly independent fragments of information. Experimental protocols are established tentatively, based on probabilities. An experiment that turns out as expected is sometimes interesting. But in general the result is much more valuable if it's a surprise. In fact, you can almost measure the significance of a scientific study by the intensity of the surprise it provokes.

This aspect of unpredictability shows up clearly throughout the history of science. Who would have said in 1850, before Pasteur and Koch, that infectious diseases would turn out to be the result of invasion by specific germs? Or in 1950, before the work of Watson and Crick, that the chemistry of heredity would be understood before that of muscle tissue? And this unpredictability holds not only for basic research but also for its applications. If during the Stone Age people had

5. Lewis Thomas, *The Medusa and the Snail* (New York: Viking, 1979), p. 73.

wanted to develop tools for cutting and slicing, they would have produced stone axes in various shapes and at various prices, but they
would never have discovered bronze. Or if, at the end of the nineteenth century, people had wanted to improve methods of locating
fragments of projectiles left in wounds, they would have produced all
kinds of probes, of all sizes, made from various materials, but they
would never have been able to predict the existence and use of X rays.

As I mentioned above, there is a category of people for whom the
unpredictable character of research is hardly tolerable, namely politicians and administrators of science, who are wary of projects that
lack a precise goal. They prefer huge programs whose aim—"objective" as we say today—is well established: the human genome, cancer,
AIDS, and so on, all efforts for which administrators reckon they can
establish plans and a timetable. But such plans can succeed only if they
are based on research already well under way, in well-established disciplines that will continue to forge ahead, using proven ideas, methods, and techniques. However, research in utterly unknown fields—
that is, in emerging sciences, hesitant and stumbling, where problems
are still ill-defined and the data ambiguous—has nothing to do with
carefully laid plans. There is little chance of finding solutions to half-
formulated questions. This kind of research has an unbridled, delir-
ious, almost wild aspect that is difficult for the public to understand.
How do you establish long-term plans in such turbulent conditions?
How do you evaluate the stages of a development whose outcome is
totally unpredictable?

What politicians can do, what they must do, is to determine the
importance of science for the country and define the share of the
budget to be allocated to it. What administrators can do, what they
must do, is to determine the relative importance of the various scientific disciplines and divide up the "research" budget among them.
These two functions were illustrated very well by General de Gaulle
on his return to public life in 1958. Recognizing the importance to

France of scientific research, he first created the Délégation à la Recherche Scientifique et Technique and funded it aggressively, over the long term. Then, showing exceptional foresight, as reported by Raymond Latarjet[6] in the following anecdote, de Gaulle appointed a committee of twelve "wise men," each representing a scientific discipline, to advise him. At the end of a year he decided to choose several areas of research which, owing to their compelling interest, would receive special funding. He assembled the twelve wise men to help him decide. He had them gather round a table and gave each of them five minutes to present an area of research that seemed especially worth supporting. And so they did. Latarjet proposed "molecular biology." At the end of an hour, there was silence. De Gaulle spoke: "You might think that a general would be particularly appreciative of spectacular projects whose descriptions he understands, whose perspective he shares, and whose developments, consequences, and repercussions he readily anticipates. Examples among those I've just heard would be renewable energy resources, the conquest of space, and the exploitation of the oceans. But deep down, I wonder whether this mysterious molecular biology, of which I know nothing and will certainly never understand anything, isn't the most promising of the mid-term developments—unpredictable, rich, capable of doing much to advance our understanding of the basic phenomena of life and its disorders. Perhaps it will be the basis for a new medicine that we cannot even imagine today. It might even be the medicine of the twenty-first century." And molecular biology was the first choice of the committee. What an astonishing vision of the future! As astonishing as that of June 1940 regarding the development and the outcome of the war![7]

It is thus possible for an exceptional politician to discern the sig-

6. *Laboratoire Raymond Latarjet* (Paris: Institut Curie, 1992).

7. In June 1940 de Gaulle opposed the Franco-German armistice and fled to London, where he organized the Free French forces. [Trans.]

nificance of a new discipline and provide the means for developing it. But the very birth of molecular biology is a good illustration of the impossibility of organizing research in a new area, of "planning" it, as we say. Molecular biology aims to explain the amazing properties of living creatures—the same ones that not so long ago seemed to require the intervention of a "vital force"—as the result of the structure and interactions of the molecules that make up the organisms. This branch of biology was born of individual decisions taken by a small number of scientists between the end of the 1930s and the beginning of the 1950s. These scientists came from highly varied backgrounds—biology, physics, medicine, microbiology, chemistry, crystallography, and so on. Recognizing that questions raised by genetics are at the heart of the study of the living world, they invented a new biology. No one pushed them in that direction. No administrator, funding agency, or national science adviser steered them down that path. Rather, it was their individual curiosity—their desire to find new ways of looking at old problems—that led these few men and women to resolve the question of heredity. The history of molecular biology provides a model for understanding how original research takes shape, independent of potential applications. Those came only later, with the possibility of manipulating the genes of organisms; the techniques we now call genetic engineering.

Here again, the conditions necessary for the birth of genetic engineering arose in an almost totally unpredictable way. During the 1950s, researchers who worked with bacteriophages, viruses that attack bacteria, observed a strange phenomenon. A certain virus was capable of reproducing itself in two strains of bacteria, A and B, when it had been prepared with strain A. When it had been prepared with strain B, however, it could multiply in strain B, but no longer in strain A. Once the oddity had been noted, many researchers lost interest in it, with the exception of two Swiss biologists, Jean Weiglé and Werner Arber. Weiglé died soon after he had begun working on this problem. Arber pursued his research in a climate of almost total indifference, notably that of the committees and organizations tasked with allocating funds for research. Yet he stubbornly carried on. In a few years he was able to show that the phenomenon was due to the presence, in certain bacterial strains, of enzymes whose function was to cleave foreign DNA and prevent it from invading the bacteria and taking root there. These enzymes are extremely specific. Each recognizes a particular short sequence of DNA. Each constitutes a veritable pair of genetic scissors that can be used to cut DNA into precise fragments for detailed study. Who would have thought that the study of the phenomenon discovered by Weiglé and Arber would lead to the surprising development of genetic engineering?

Sometimes predictions made at a given moment are overturned by new developments in research. An example is the idea of so-called cloning of human beings. In talking about microorganisms, the term "clone" designates all of the genetically identical individuals produced by the divisions of a single cell, that is, the result of asexual reproduction. In complex, sexually reproducing organisms, scientists wondered for a long time about the respective roles of the nucleus and the cytoplasm in cellular differentiation. In particular, they wondered whether the nucleus, which in the egg is capable of giving rise to all tissue types, loses some of this potential after differentiation. To answer the question, scientists conducted experiments in "nuclear transplantation," replacing the nucleus of an egg cell with a nucleus from a differentiated cell—intestine, kidney, and so on. Work with frogs had shown that these transplants could, in certain cases, lead to development of tadpoles, even adult frogs. This finding led to the idea that, starting from cells taken from Brigitte Bardot or General de Gaulle, you could make as many Brigitte Bardots or General de Gaulles as you liked. Ten or fifteen years ago, a flurry of articles appeared describing the advantages, or more often the disastrous consequences, of this kind of cloning. In other words, many people did not hesitate to extrapolate to humans what could be achieved only painstakingly in frogs. But efforts carried out over the last fifteen years have shown that this kind of exercise doesn't always succeed, that it cannot be applied to all organisms, and is particularly difficult when it comes to mammals. Many researchers have tried with mice; no one has yet managed to clone one successfully. As soon as the fertilized egg has divided twice and the embryo consists of four cells, the nuclei of these cells become incapable of assuring the development of a complete embryo. A fortiori, the nuclei of adult cells are even less capable of doing so. On the other hand, Scottish researchers recently announced the birth of a lamb produced by inserting the nucleus of a mammary cell from a sheep into the enucleated egg of a sheep. We don't yet understand why a sheep-cell manipulation can do, at present, what a mouse cannot. And it is impossible to say whether we will ever be able to produce Albert Einsteins or Ava Gardners on demand. On this subject, people can continue to indulge their own private fantasies.

If it is difficult to predict the future, it is sometimes just as difficult to reconstruct the past. Scientific thought both refines and complicates the answer to the old question, where do we come from? The whole universe and the objects it contains, living or not, are like the products of an evolutionary process in which two kinds of factors are at play: on the one hand, the constraints that determine the rules of the game and demarcate the limits of the possible; on the other, the circumstances that direct the actual course of events, the history of the game as it was actually played. The constraints can usually be formalized. Everything that depends on them can be predicted with high probability. The role of history, on the other hand, can be recognized and sometimes explained. But obviously the chain of events that will make history tomorrow cannot be predicted. This feature of the forces that shape our world is entirely contingent.

The relative role of constraints and of history varies according to the domain. History was long absent from the world of physics, for example. None of the equations of classical physics included a parameter for time. In an immutable universe where past and future were indistinguishable, time was reversible. It was only at the beginning of this century that time came into play in physics. In the new cosmology, the universe, galaxies, stars, elements, and particles all acquired a history.

Few novelists can match the imagination of physicists when they talk about the history of the universe. Through their calculations, physicists arrive at a reality whose mathematical evidence is opposed to the data provided by sensory intuition. The birth of the universe, twelve to fifteen billion years ago, is thought to have been the consequence of fluctuations of energy in the initial void, creating a different kind of void, lacking matter but filled with energy. This situation would have led to a sudden expansion, an explosion, in other words, the Big Bang. Such fluctuations of energy would have been very rare, but it is not unthinkable for similar events to have occurred in other regions of space, in other times, leading to the creation of other universes. According to these theories, our universe would not necessarily be the only one. It would not be the center and the stage for everything that happens in the world. Its beginning would perhaps not have been the beginning of everything.

As physicists tell the tale, a thousand-billionth of a second after the big explosion, when the temperature of the universe "fell" to a million billion degrees, particles and antiparticles were rapidly created and annihilated. Then, with the expansion and cooling of the universe, annihilation began to outpace creation. Almost all the particles disappeared. And if there had not been a slight excess of electrons over antielectrons and of quarks over antiquarks, ordinary particles, those that form the very basis of matter, would today be absent from the universe. It was this slight excess of matter over antimatter—an excess estimated at one ten-billionth—that survived to form, three minutes later, the light atomic nuclei; then, after a million years, the atoms; much later, the heavy elements in the stars; and finally, the stuff out of which the living world arose. If there had not been that excess of a ten-billionth of some particles over other particles, our universe would not exist, at least not in the form in which we know it.

As for the earth, its composition is thought to be a by-product of the formation of energy in the stars and successive waves of births and disappearances of stars in our galaxy. Today it is estimated that, four and a half billion years ago, the earth formed by accretion: cosmic dust agglomerated into granules, the granules into gravel that became little stones, then rocks, and small planets. Finally the dust became the size of the moon, then the earth. Amazing scenario. This earth on which we move, which supports the oceans, the continents, and the mountains, and which is inhabited by myriads and myriads of living creatures—this earth would have been formed by the progressive agglomeration of dust from the heavens. And what about the conditions necessary for life that don't exist on other planets, like Mars and Venus—water; the right distance from the sun, which keeps things from getting too hot or too cold; and so on? How many thousands of events, totally independent, any one of which might not have happened, had to have occurred in a certain order to create the universe, our galaxy, our solar system, and the earth? For all of this history, science can explain how this or that event occurred, after the fact. But it can never predict.

In trying to explain the origin of life, biologists for their part must call on all their imaginative resources. It is obvious that, for the living world and its evolution, the role of history is of the utmost importance. Life seems to have appeared fairly quickly, probably less than a billion years after the formation of the earth, in the form of something we might call a "protobacterium." Life means reproduction. But the apparatus of reproduction that we observe today in the simplest organism, the humblest bacterium, already features a formidable complexity. The duplication of DNA alone brings into play an enor-

mous number of proteins, the synthesis of any one of which demands
an even more considerable number and diversity of macromolecules.
It is unthinkable that such a system would have emerged fully formed
from Zeus's head; hence it is necessary to imagine more or less plau-
sible scenarios that might account for a progressive buildup of complexity. According to one scenario, which has become fashionable
over the past several years, the living world as we know it, dominated
by DNA, was preceded by a world in which RNA functioned both as
a catalyst and in replication. It goes without saying that the emergence
of this RNA world and the transition to a DNA world imply an im-
pressive number of stages, each more improbable than the previous
one. Moreover, most of the hypotheses required for such scenarios
lend themselves neither to reconstruction nor to experimental veri-
fication. In other words, although it seems clear that humans, animals,
plants, fungi, and microbes—in short, we living beings—are all descended from an initial protobacterium, we are not even close to
knowing the true face of our common ancestor.

In the same way that the theory of evolution changed the idea of life
on earth, modern cosmogony has replaced the idea of an unalterable
universe and reversible time by that of a universe in perpetual flux,
abandoned to history. The universe, like life, has a beginning. Like life,
it evolves. Yet our senses and our brains were not selected to perceive
the properties of electrons or the distances that separate galaxies or
cosmological time, but to deal with the world around us—objects,
space, and time—on the scale of human beings. To imagine what
came before or what will happen after, we have to trick ourselves. And
it is not certain that we will ever manage to reconstruct what really
happened. As Claude Lévi-Strauss[8] pointed out with obvious satisfaction, the accounts that science ultimately resorts to are as removed

8. *Histoire de Lynx* (Paris: Plon, 1991), pp. 11–12 (English trans., *The Story of Lynx* [Chicago: University of Chicago Press, 1995]).

from common sense as the products of mythological thought. When we think about the origin of life, we have to accept that, over the course of some eight or nine hundred million years, thousands of events, each highly improbable, followed one after the other to permit the transformation of an earth without life to life in an RNA world, and then to life in a DNA world. Clearly, such a history might appear as incomprehensible to noninitiates as do the stories of Creation in the *Theogony* of Hesiod, the Upanishads, or the Bible. Indeed, mythological tales seem closer to common sense than does the discourse of biochemists and molecular biologists.

The molecular biologists, faced with a difficult problem they are unlikely to solve for a long time, have recourse to three possible solutions. Some biologists, including some of the greatest, consider the appearance of life on earth so improbable that they prefer, half jokingly, half seriously, to invoke a kind of panspermia. The seeds of life would have arrived on earth aboard a spaceship sent from a faraway planet by a civilization more evolved than our own! Which, of course, only reduces the problem a notch. This solution is the least common.

Others consider that the appearance of life on earth was so improbable that almost certainly it happened only once. It resulted from such an unlikely series of events—any one of which might not have happened—that there might well never have been life on earth. These same scientists also tend to believe that very probably no other conscious life exists in the universe.

A third group of scientists has an entirely different attitude. For them, the stages implied by the advent of an RNA world and passage to a DNA world were the result of ordinary chemical reactions that could not help but occur given sufficient opportunity, that is, time. These scientists reason that it would be impossible for life *not* to have formed on earth. Besides, sensitive to the argument of astrophysicists that the universe contains numerous planets with conditions apparently similar to earth's, they maintain that there must be many places in the universe that support life, probably even conscious life.

On the basis of current knowledge, the choice between these last two options is above all a question of taste. Some prefer to cultivate the idea that life is restricted to earth and, consequently, that the ability of human consciousness to reflect on the universe and what is in it is unique. Others, by contrast, prefer to believe that life is commonplace and that if it exists on other planets, its qualities could not possibly be very different from those observed on earth. Convinced, furthermore, that once begun, life must necessarily lead to consciousness, this last group strives to find means of communicating with the other civilizations that, according to them, must surely occupy other regions of the universe.

Until now, however, no one has managed to detect a trace of a signal suggesting life and coming from the galaxy or beyond the galaxy. Recently, our attention was drawn to a meteorite that *might* have come from Mars and that *might* contain a structure bearing some resemblance to that of the oldest living structures found on earth. But the arguments put forth are hardly convincing. This story seems to have had more to do with the publicity related to NASA's space flights to Mars than with reality. Everything we learn about even the most varied organisms living on earth shows that, in all likelihood, they are all descended from one and the same ancestor. Thus it does indeed seem to be the case that life appeared once and only once on earth; that it resulted from a series of events, each highly improbable; and that if any of these events had not occurred, life as we know it would not exist.

To get to the Pasteur Institute in the morning, I walk through the Luxembourg gardens. Each year, entering the gardens on a spring day, I feel the same shock, the same amazement. Each year I marvel again at the buds bursting and beginning to open, at the green lace of nascent leaves that hangs on the branches and trembles in the breeze as if it feared not being able to carry it off. But what is amazing is that it does carry it off. Once again, the system works. Once again, the days will grow longer, the light and the warmth return. The leaves will form, then the flowers, then the seeds. Animals and plants will burst with life and growth. Not the slightest hitch, not the least little error. The program is fixed. Indifferent to the affairs of men, the great machine of the universe will continue to function inexorably and impeccably. More than the oceans and their storms, more than the mountains and their glaciers, more than the canopy of the heavens and its galaxies, the return of this little green quivering that runs through the trees and surprises me on a spring morning gives me the feeling that I am witness to the awesome show that, for some twelve billion years, has enlivened the great stage of the universe.

Physicists can explain how matter forms and how the forces that act on it behave. But I still have not figured out whether they can conceive of a universe endowed with other properties. That is, whether in the mingling of constraints and history that shaped the universe, only history entailed an element of contingency, or whether, in the beginning, the constraints—what we call the laws of nature—could also have been the result of contingency.

As a child, I believed with all my heart in fairy tales. For me, they described a view of the world as real as anything I saw on the street or in the countryside. Ogres and giants, who, as we all knew, eat little children, seemed to me hardly different from certain people I'd pass in the gardens and of whom I was told to beware. The transformation of Cinderella's pumpkin into a carriage struck me as no more extraordinary than the card tricks my father would do when he was in a good mood. None of these things led me to doubt. Here or there, the world was that way. Things just were like that.

That same feeling still comes over me every spring with the reappearance of the leaves. Over time some of my convictions have altered. My faith in the power of fairies and the force of ogres has diminished a little. But when I see these festoons of leaves that reappear so faithfully every year, I am bowled over by the gratuitous nature of the world that surrounds us. In other words, if in the sequence of events that generated the elements, the galaxies, the stars, and the earth certain events had not occurred, or had occurred at the wrong moment, there might be no leaves on the trees—or perhaps no trees—perhaps even no living world. How can we not see our world and the way it works as arbitrary, even whimsical? Things just are the way they are, especially in the living world. How can we not wonder at oddities like death and aging? What possible necessity can there be for trees to bear leaves that, in many cases, drop every fall only to grow anew every spring? Or for animals to have four legs? Or for living creatures to reproduce, in most cases requiring two to make a third? Or for reproduction to be the one bodily function assured by an organ individuals only ever possess half of, which obliges them to devote a lot of time and energy looking for another half? Yet that's how things are. Pointless to wonder whether the world might have been different, or whether it might not have existed at all. For the scientist, this world has a unique attribute: it has existed and functioned for some ten or twelve billion years.

In René Clair's film *C'est arrivé demain* (It happened tomorrow), a young reporter meets a ghost and gets into his good graces. Every day the ghost sends him the next day's newspaper, which gives the young reporter unrivaled power. He knows what tomorrow will bring. He is on top of events, risks, schemes, racing results, and the ups and downs of the stock exchange. Naturally, he succeeds in everything he undertakes, including love. Everything goes well, until the day the newspaper announces that the reporter will die the following day. Panicked, he runs away to avoid his destiny. But no matter what he does, he cannot escape it. He finds himself at the hour and place predicted for his death. And if, in spite of everything, the film has a happy ending, it's only because newspapers don't always get their stories right.

2　The Fly

....

"*Because without flies, no fly swatters, without fly swatters no
governor of Algiers, no consul, no insult to avenge, no olive trees,
no Algiers, no great heat, gentlemen, and besides, the great heat
is the health of tourists.*"

—JACQUES PRÉVERT

Brno, 7 August 1965. A crowd jostles around the cathedral, making
a tremendous crush in front of the doors, which the police guard im-
passively. For the first time in twenty years, a mass is going to be celebrated here. Yet despite their thirst for religion, the ordinary inhabitants of the city will not be allowed in. This mass has been organized
by the Academy of Sciences of the Democratic Republic of Czecho-
slovakia to commemorate a very special occasion, the one-hundredth
anniversary of the first account by the monk Gregor Mendel on heredity in garden peas. The event would simultaneously celebrate one
of the great men of Czech science and what is generally regarded as
the official birth of genetics.

The Czech Academy invited geneticists from all over the world to participate in a symposium to discuss both Mendel's role in genetics and the developments in that discipline during the first half of this century. Since Mendel was a man of the church, and since he had himself celebrated mass at this site, the Czech Academy decided after much wavering to end the symposium with a high mass in the memory of its hero: a mass limited to the participants of the symposium. The result is an extraordinary scene in the nave. On one side a hundred American geneticists, delighted by the occasion but flabbergasted at finding themselves attending it in a Communist country; on the other, a hundred Soviet biologists, faces expressionless, arms folded, bewildered at finding themselves at mass. Scattered among them a hundred European and Czechoslovak geneticists perplexed by their position between the American bloc and the Soviet bloc. And, as if moved by a single impulse, all these people surge to their feet as the vault resounds with trumpets accompanying a Bach cantata, and the Bishop of Brno and his entourage move slowly forward.

Clearly, Prague Spring had arrived.

Throughout this century, genetics has been mired in politics. In any country, a symposium organized to commemorate the hundredth anniversary of this field and its founder would seem the most ordinary of tributes. But at the time and in a socialist republic, things were different. Such a gathering could only be permitted once genetics had been rehabilitated and its main opponents thwarted. For just as the Church had once condemned the ideas of Galileo as incompatible with doctrine, the Communists had forbidden genetics as incompatible with the principles of Marxism.

This ostracism began at the end of the 1920s, in Soviet Russia. In the name of dialectics, the neo-Lamarckian Communists, who supported the idea of the inheritance of acquired characters, had attacked the Russian geneticists, not with scientific arguments, but with texts

written by Engels. The leader of this affair was an agronomist named
Trofim Lysenko. In the early 1930s he had made a name for himself
by what he called his "discovery," a method for planting wheat in summer and harvesting it in winter. This method had nothing original
about it and was rapidly abandoned. But Lysenko maintained that his
findings, which he considered a major success, could not be explained
by genetics. Without any hesitation, he concocted an alternative theory, for which he had no experimental support, based on the inheritance of acquired characters. He then launched an assault on the Soviet school of genetics, which up to that point had been brilliant. This
charlatan was paranoid.

The style of Lysenko's declarations, like the substance, betrays his
total incompetence, his ignorance not only of elementary biology but
of the scientific process itself. As Jacques Monod[1] pointed out, they
bring to mind the ramblings of self-published autodidacts convinced
they have found the secret of life or a cure for cancer and furious because "official science" is ignoring them. But what is most disconon-
certing—what really proves Lysenko's cleverness—is that he succeeded in obtaining the support of Stalin and, along with it, of the
entire Soviet apparatus—the state, the party, the courts, and the
press. With this support Lysenko was able to defeat his enemies. Both
the practice and teaching of genetics became forbidden. Those who
refused to subscribe to his theories were sent to Siberia, and many
never came back. Naturally, what was true of the Soviet Union was
true as well of the countries it dominated. Thus genetics was subsequently banned in all the socialist republics. At the University of Sze-
ged in Hungary, the rector led a parade of the full faculty, who sol-
emnly carried the collections of fruit flies used for research and
teaching to the latrines for disposal. At Brno the statue of the "monk

1. Jacques Monod, foreword to Jaurès Medvedev, *Grandeur et chute de Lyssenko*
   (Paris: Gallimard, 1971).

Mendel" was unbolted and torn down. As for the pea plants that Mendel had used in his experiments, they were ripped out of the monastery garden. Let's not refute, let's just destroy!

Most amazing in this affair were the arguments of Lysenko and his followers. When debating with Soviet geneticists, Lysenko never took into account the data from experimental science, the countless results about heredity in animals and plants accumulated during more than thirty years, thanks to genetic analysis conducted in various countries. The surprising successes in agriculture on which he prided himself were contested by legitimate biologists and in no way justified his violent diatribes against chromosomes. When Lysenko spoke about biology, what he said was so ridiculous that he immediately destroyed any credence anyone might have given to his agricultural pretensions. At least that was the conclusion of Boris Ephrussi, the first professor of genetics in France. Ephrussi, of Russian origin, had had the opportunity to speak at length with Lysenko, whom he described as a stubborn man who uttered, without batting an eyelash, such inanities as "There are two kinds of glucose, plant glucose and animal glucose"; or "Amino acids are primarily involved in osmotic equilibrium in cells"; or even "The cytoplasm is what gives the cell's nucleus its properties"—all propositions for which there was no shred of evidence.

For Lysenko and his supporters, the concept of species was just a bourgeois notion. They didn't hesitate to publicize loudly experiments that claimed to transform one species into another—wheat into rye, oats into barley, cabbages into rutabagas, then all three into fir trees. These operations were intended to testify to the success of progressive science. For Lysenko, however, the real debate was actually not scientific but ideological. The argument he used endlessly against genetics was its incompatibility with dialectical materialism. For him that was the real stakes, the heart of the matter, the sole basis on which he could obtain the support of Stalin and the Soviet government. It was only by wielding this argument that he had a chance of winning, of having his ideas prevail, of putting down his enemies. And in fact Lysenko was absolutely right. No matter how you approach it, it is impossible to fit genetics to dialectics. Genetic theory could no more mesh with Engels's *Dialectics of Nature* than could Darwinian theory—the theory of natural selection—which Lysenko also rejected. For him, only the inheritance of acquired characters, whose validity he thought he had demonstrated, allowed nature to be modified in any lasting way, and thus made to conform to Marxist doctrine. So Stalin had no hesitation in backing Lysenko.

It is easy enough to understand how, under the pressure of ideological terror and police dictatorship, Russian biologists had to yield to or even embrace Lysenko's arguments. But what about those in the West? What about those who, having no reason to fear for their lives, their liberty, or their livelihood, supported this web of lies and contradictions with such enthusiasm? How can we forget the frenzy of a certain segment of the press and leftist intellectuals who, blinded by passion and closeted in their ideology, were driven to forsake reason and to stoop this low? What explains the sudden interest in biology among men previously proud of their ignorance of or even scorn for science, certain as they were of the incontestable superiority of culture over nature?

I had the opposite reaction to the Lysenko affair. At the end of World War II, when I was just beginning to surface, it was in full swing. It helped to steer me toward science, and more precisely toward genetics. I was amazed to discover that in the middle of the twentieth century it was possible for a charlatan to obtain the support of the government in his country to impose, on the one hand, a silly "scientific" theory and, on the other, a disastrous agricultural method. It was possible for an unscrupulous person to target, with intent to destroy, one of the most solidly established scientific theories. Nothing prevented a political dictatorship from imprisoning sci-

entists accused of practicing "bourgeois science" in the service of re-
actionary politics. Above all, perhaps with even more astonishment,
I discovered that it was possible for men as free as were our intellectuals to line up behind Louis Aragon and, filled with ideological passion, to yield to slavishness and degradation. To me, at the time,
studying genetics meant refusing to substitute intolerance and fanat-
icism for reason.

Lysenko was in office a long time—long enough to destroy Russian
biology and agriculture, which never fully recovered. He survived
Stalin. But when he came under attack by Soviet physicists, he disappeared at the beginning of the 1960s. Because of nuclear weapons,
physicists had the opportunity to travel abroad and attend confer-
ences, unlike their colleagues in biology. The emergence of molecular
biology had not failed to interest Western physicists, several of whom
even played leading roles in its development. Mixing with them, the
Russian physicists quickly realized how stupid and dangerous Lysenko's theories were. They managed to convince the Soviet author-
ities of it. After some ups and downs, Lysenko was eventually stripped
of his titles and his authority. At Brno, the statue of Mendel was put
back on its pedestal. The peas were replanted in the garden. The cen-
tenary symposium could take place.

After peas, genetics turned its attention to flies, which serve not only
as favorite objects of geneticists but also to amuse children. I was introduced to the joy of the fly by a little school chum. It happened toward the end of the third grade. At that time, one of our classmates
went around making a particularly miserable impression: pale, gaunt,
knock-kneed, always wearing clothes that were too small for him, Antoine aroused your sympathy the instant you laid eyes on him. He had
been orphaned from infancy and lived with his aunt, a real shrew to
hear him tell it. She was bad-tempered, malicious, and stern. She used
to strap his legs with a belt for every grade she judged not good enough

or for any remark she thought impertinent. Antoine's descriptions of his aunt gave me nightmares.

He had one consolation, Antoine did, and that was flies. We became good friends. He trusted me. At recess, he would often pull me into a corner of the playground to show me his flies. From his pocket he'd take out a fly cage made from two rounds of cork connected by a series of pins that formed bars. In the cage you could see a few emaciated flies dragging themselves from bar to bar. On certain days he considered special, Antoine would play what he called "the big game." He would raise one of the pins and fish a fly out of the cage with his skinny, dirty fingers. "We're going to try to understand how a fly is built," he would say. And he would begin to take the fly apart. He'd pluck the legs out, one by one, like single hairs, leaving only the body and wings. To dismantle these, he proceeded differently. He'd tug the wings off gently, to "unfasten" them, he said, without tearing them. Afterward he'd want to "reassemble the system" and put everything back in place. But of course it wasn't possible. All we could do was watch the fly struggle, observe its quivering subside little by little into stillness.

"To understand how a fly is built" has been the ambition of geneticists throughout this century. It was an American, Thomas Hunt Morgan, who first promoted the fruit fly to its privileged status in work on heredity. Morgan was an embryologist. For a long time the study of embryonic development had been limited to morphological examination. Changes in the form of an embryo and differential cell multiplications in various regions of the body were observed under a microscope. At the end of the nineteenth century, experimental studies of these phenomena were just beginning. But the so-called "mechanics of development"—the forces that permit a structure as minimally organized as an egg to transform itself into an organization as complex as an animal—remained beyond the reach of experimentation. So much so that to account for this extraordinary metamorphosis, which seemed beyond any possibility of comprehension, some embryologists came up with the idea of a sort of vital force, a "principle of development" that had nothing to do with the laws of physics. For Morgan and his friend E. B. Wilson, however, heredity held the key to the mechanism by which a chicken forms from a hen's egg, and a human being from a human egg. Hence Morgan's decision to give priority to the study of heredity. At first he worked on mice and rats, but he very quickly abandoned them because mammals were too expensive, reproduced too slowly, and were prone to infection. So Morgan turned to the fruit fly *Drosophila*.

Nothing predisposed the fly a priori to a glorious destiny in science. Aristotle mentions a gnat that emerged from the residue of vinegar, most likely a relative of *Drosophila*. The genus was described and named at the beginning of the nineteenth century. Unfortunately, perhaps, the name *Drosophila*, or lover of dew, won out over "Oenopote," or drinker of wine, which had been used by several entomologists. The best-known species, *Drosophila melanogaster,* was described in the middle of the century. It appears to have originated in the tropics. Most likely it was introduced to Europe and the United States in cargoes of bananas.

This little fruit fly made its appearance in a laboratory at Harvard University at the very beginning of this century. It quickly demonstrated remarkable qualities: it was small, easy to raise in the laboratory, able to withstand mutation and crossbreeding experiments, fertile all year without interruption, and prolific, producing a new generation every twelve days, or some thirty generations per year, with each female producing close to a thousand eggs; males and females were easy to distinguish; it had only four chromosomes. In short, it was the ideal organism for studying heredity.

At the time, Morgan was working at Columbia University in New York. A colleague in a neighboring laboratory drew his attention to the virtues of the fruit fly. The first specimens arrived in the laboratory at Columbia in 1907, brought by a student of Morgan's. Morgan himself became interested in the summer of 1909. New types appeared very quickly in the population, in particular, a male with white eyes, not red like those of its congeners. This event unleashed an avalanche of findings. The fly's descendants included individuals with white eyes—a mutation. But the "white eye" characteristic was inherited in a very special way. Only a fraction of the males had white eyes, while the females all had red eyes. This suggested a recessive trait linked to sex. Other types of mutants appeared, several of which were sex-linked, indicating that the genes governing these traits were carried on the X chromosome. In another few weeks, Morgan and a student demonstrated that chromosomal fragments were exchanged among homologous chromosomes and recombined. They showed that the amount of recombination is an indication of the distance between genes on the chromosome, and established the basis for gene mapping on chromosomes using this technique.

Morgan knew then that the fly would allow him to decode heredity. He took on several particularly brilliant students—C. B. Bridges, A. H. Sturtevant, and H. J. Muller—whom he set up in a laboratory known thereafter as the "fly room." In the fly room, Morgan and his team worked wonders.

The fly room was quite small, and crammed with tables, desks, microscopes, and bottles used for raising the flies. It was home to a dozen researchers, students, postdoctoral fellows, and visitors. Ideas shot out of the fly room like rockets, one experiment following another in an endless series of discoveries, discussions, and theories. Morgan and his group worked daily in close cooperation in an atmosphere of excitement and enthusiasm mixed with a sense of critical inquiry, generosity, and open-mindedness. Their story remains one of the rare great sagas in the history of biology. Open exchange was a permanent feature of the fly room. As new findings emerged or new ideas came up, the group discussed them freely, to the point where everyone forgot who had first generated an idea. This way of working helped speed up the discoveries.

Hundreds of thousands of flies were cultivated by Morgan and his group, yielding a continuous stream of new mutants. In a few years, the principal features of heredity, what would become the "laws" of genetics, were elucidated. As Ernst Mayr[2] points out, while all the geneticists of the preceding century and the beginning of this century had failed to find the right answers for want of asking the right questions, Morgan succeeded brilliantly. Rather than asking questions about the physiology and chemistry of genes or speculating about possible theories of heredity, he stuck to the facts, thereby founding a genetics that interpreted Mendelian inheritance in terms of chromosomal theory.

The activity in the fly room at Columbia continued for close to twenty years. From there it spread to most laboratories and universities worldwide, and mutants began to pile up. Little by little, the laws of classical genetics and chromosomal mechanics were refined. The image upon which genetics was built in the 1930s was that of genes threaded along the chromosomes one after the other, "like beads on a string."

Once Morgan and his colleagues had established that genes were carried on chromosomes, biology was open to a whole series of new questions. What is the chemical nature of genes? What is the hereditary material? How do genes duplicate? How do they serve to determine the properties of cells and organisms?

A first indication of the function of genes came from medicine, from ideas that would lead to the concept of genetic diseases. In 1902,

2. Ernst Mayr, *The Growth of Biological Thought* (Cambridge, Mass: Harvard University Press, 1982), p. 754.

the English physician Archibald Garrod had become interested in al-
kaptonuria, an ailment marked by black urine. According to Garrod,
this disease was the result of what he called an "inborn error of metabolism," that is, a congenital deficiency that involved the interruption of a metabolic chain that ended in the formation of urea.

But to go more deeply into this idea, to answer all the questions
raised about genes, the fly was hardly the right tool. Although it was
possible to scrutinize the chromosomes under the microscope—especially the giant chromosomes of the salivary glands of the fly—and
to pick out bands of unequal size, there was no way to establish any
correlation between the bands and the functions attributed to the
genes. And though one could graft tissues fated to become specific
organs, such as eyes, from mutant flies into normal flies (and vice
versa) and see how those organs developed, as Boris Ephrussi and
George Beadle had done, the results were indirect and not very convincing. Until World War II, ideas about the structure of the gene
were no more than hypotheses based on data obtained from the study
of mutations and the effects of mutagenic agents like X rays.

At the end of the 1930s, the fly was relegated to the property room.
Microorganisms—fungi, yeasts, bacteria, and viruses—took its place
and maintained their priority for close to thirty years. As a matter of
fact, it was thanks to bacteria that the role of deoxyribonucleic acid
(DNA) as the hereditary substance of the genes was identified; that the
relationship between genes and proteins was defined, the mechanism
of synthesis of proteins clarified, the genetic code solved, and the process of gene replication elucidated. In short, it was through bacteria
that the nature and function of genes were established.

It then became possible to analyze the genetic structure of a bacterial cell as well as of some of its functions. In particular, with Jacques
Monod, we showed that genes do not all function continuously in the
cell. Many of them carry out their activity only upon request, in response to signals from the environment and to the nutritional needs

of the bacteria. There are regulatory circuits, themselves governed by specific genes, that modulate the expression of many genes.

Although the new molecular genetics allowed access to the innermost depths of the bacterial cell, complex organisms remained beyond reach. They could still only be studied using the methods of classical analysis, which involve crossbreeding organisms that differ in several characters, or using a new technique: fusing cells together, which made it possible to localize certain genes on chromosomes and to study a few functions. But the DNA of these organisms is so complex that it resisted the methods of molecular genetics for a long time.

In the summer of 1972, however, an especially interesting meeting inspired by Ernst Hadorn was held in Zurich. Hadorn was an embryologist. He had started out working with amphibian embryos, but faced with the difficulty of analyzing genetic factors involved in the development of these organisms, he switched to *Drosophila*. Hadorn studied the differentiation of "imaginal discs," groups of cells that, in the larva, prepare the formation of the adult fly. The adult fly is actually built like an automobile: there is a "disc" for producing each eye; one for producing each wing; one for each leg, and so on. The elements are thus prepared separately, then assembled at the end. This differentiation is clearly under the control of the genes. In trying to bridge the distance separating the genetic study of *Drosophila* from that of bacteria, Hadorn had invited fifteen *Drosophila* experts and fifteen molecular biologists. Each told his story as simply as possible, so he could be understood by the others. But an enormous chasm remained between the two groups. There was as yet almost no way to see how the powerful techniques used in research on bacteria could be applied to flies.

A few years later, the landscape was completely transformed. Little by little, we learned to manipulate DNA in bacteria. We soon succeeded in cutting the long strands into specific fragments at precise locations, in joining the fragments end to end, and in inserting the segments into a chromosome. All these manipulations fell under the rubric of genetic engineering. Gradually, it became possible to manipulate the enormous quantities of DNA contained in complex organisms, to pick out certain genes, isolate them, make numerous copies, and do a detailed analysis of their structure. It was also possible to take a gene from one organism, insert it into another, and examine its functioning. From that point on, the gap that separated bacteria and complex organisms began to close. Genetic analysis could be extended to any organism, including humans. The methods of molecular genetics even made it possible to circumvent the constraints of classical genetics, that is, to avoid crosses between brothers and sisters or between parents and children (a technique in any case not applicable to humans). The field of human genetics could now move forward.

The fly acquired a new scientific life. Once again it became the organism of choice for studying one of the most daunting problems in biology: the development of the embryo, especially the role of genes in development. The very fact that dogs come from dogs and wheat from wheat proves that reproduction and embryonic development are governed by heredity, that is, by genes. But what was not understood was how genes determined the form of the new organism in the course of its development; how from one cell—the fertilized egg—billions of cells all containing the same genes could form and differentiate into structures as varied as the cells of nerves, bones, muscles, the intestines, and so on.

Until the 1970s, these questions remained beyond the reach of experimental biology. But the ability to isolate a gene detected in an organism, purify it, sequence it, and pin down its function opened new doors to the study of embryonic development. The old question, how does heredity function in the fruit fly? was replaced by the question, how is a fruit fly built?—the same question my buddy at the Carnot school used to ask. Until then, researchers had been collecting mutations as indicators, as a way of marking chromosomes so they could study chromosomal mechanics. But now mutations became valued for opening the way to understand the role of genes in the formation of an animal. If a mutation involved a developmental anomaly, such as a deformed fly or larva, it was because the mutated gene played a specific role at some stage of development, because it directed an event essential to complete that stage. The entire old arsenal of *Drosophila* mutations was thus reconsidered from a new angle. For example, in 1916 an astonishing *Drosophila* mutant was identified: a fly that had two pairs of wings instead of one. At the time, apart from its off-the-wall and intriguing character, this mutation represented scarcely more than a point located on a chromosome on the genetic map of *Drosophila*. In the 1970s, on the other hand, the significance of this mutation lay in the morphological transformation it caused in the fly, giving rise to the question, how could a mutation change one segment of the body? How could it put wings in the place of the little balancers, or halteres, that are found on the third thoracic segment? What this mutation that produced a four-winged monster showed was that the form of an animal is dictated by the genes down to the smallest detail.

More monstrous still was another mutant identified in 1970. A normal fly has a sort of antenna above each eye. One day, a mutant appeared in a population with a leg in the place of its antenna! This second mutant confirmed the conclusions drawn from the first: that it is indeed the genes that dictate the development of an animal. Moreover, these two mutants illustrated an important idea: that the most common genetic anomalies do not involve the formation of new, previously unknown structures. In these two mutants, the anomaly was that an organ appeared where it wasn't expected: in the first case, a pair of wings where the halteres should have been, and in the second, a leg where an antenna should be. It was as if in any one area of the body all the ingredients needed to form another area were already there, ready to appear on cue.

Accordingly, the way opened to study the genetic system underlying the development of the embryo in the fly. The problem now was to isolate mutations that modify not the building materials of the body but rather the elements that order these materials and put them in place, thus defining the form of the animal. A mutation is any type of change in the genetic information, that is, in the string of bases that constitute the genetic text: substitution of one letter by another, addition or loss of letters, break in the sequence, inversion, transposition, insertion, and so on—in short, any of the kinds of errors that might occur in printing. Once they have occurred, these errors perpetuate themselves in succeeding generations. Spontaneous mutations arise only rarely. They can be induced, however, by various procedures, such as exposure to radiation, chemical treatment, and so on. The difficulty lies in detecting mutations, which are rare events, in necessarily large populations. So to obtain mutations means to increase their frequency; then it is crucial to spot them, and finally, if possible, to select them. Harvesting a desired mutation entails, first, imagining its probable effects. It is mainly a matter of shrewdness and patience.

Shrewdness and patience, not wanting among fruit fly geneticists, are notably present in the work of Christiane Nüsslein-Volhard and Eric Wieschaus. Once they had defined their target, the development of the embryo, they managed to collect what they were looking for in a few months. They were able then to analyze the process that, from the first stages of embryonic life, establishes the blueprint for the future animal. As might be expected, this architecture is built up in stages, each stage directed by a group of specific genes.

One of the questions that has long bedeviled embryologists is, what role do the maternal genes play in the formation of the embryo before fertilization? Or to put it better: does the egg as formed by the mother, which will constitute the embryo after fertilization, contain cytoplasmic elements from the mother that carry genetic information necessary for development of the embryo, over and above the interplay of embryonic chromosomes? The answer, which has come only recently, is "yes." Among the stages governing the formation of the embryo, the first ones are directed by the cytoplasmic products of the maternal genes. Only later are the genes of the embryo itself activated to direct the subsequent stages.

This miracle of nature, the formation of an animal from a cell—the fertilized egg—was for a long time one of the last refuges of vitalism. It is now described chemically in terms of the structure and interaction of the molecules involved in these processes. The first stages, those directed by the products of the maternal genes as the ovum gradually forms, establish axes that will orient the embryo by defining the front and back, and top and bottom. In this way, a system of coordinates is put into place that will allow the cells formed by the embryo to locate their position and define their identity. This initial organization is settled before fertilization.

Once these cues are in place, the genes of the embryo take control of its fate and gradual development. Under the direction of the embryonic genes, the body of the embryo, represented by the fertilized egg, divides itself up into roughly ten segments. These segments, up to that point indistinguishable, begin to differentiate and to acquire a particular identity. One can begin to distinguish the future head with eyes, mandibles, and antennae; the future thorax with a pair of wings and three pairs of legs; and the future abdomen, ending in the anal region. At last, we are in a position to observe how the body of the future animal is gradually sketched out in series of strictly geared events.

Thanks to the methods of molecular biology, it has been possible to isolate most of these genes, study their structure in detail, compare them, identify the time and site of their expression in the embryo, and sketch out the final product. This is how the existence of one type of genes that define the form of the organism was demonstrated. These genes are expressed in groups of well-defined cells that we know how to identify. They control specific events in the development of the embryo. When the structures of these genes, that is, the sequence of their bases, are compared, they turn out to be formed by particular linkages of relatively small fragments of DNA that are found in many genes. This suggests the idea that most of these genes are constructed from a limited number of genetic segments that can be meshed together. Once again, the complexity in nature stems from a combinatorial system consisting of a small number of elements.

This combinatorial system takes on new meaning when we examine the products of these genes, the proteins that control embryonic development. Each of these DNA fragments produces a polypeptide chain that corresponds to a domain, a protein motif whose three-dimensional form and electrostatic charge control its capability for molecular recognition and interaction. All the genes that direct events in the first stages of the embryonic life of a fly affect the expression of other genes by activating or inhibiting the transcription of the corresponding segment of DNA into RNA. Regulatory proteins, the products of these genes, all possess as part of their structure a domain capable of specifically recognizing a region of DNA that controls the activity of a neighboring gene. Thus, each possesses a domain that has a strong affinity for a segment of DNA. (There are a limited number of these domains, with odd names like "homeobox," "zinc fingers," "POU," and so on.) This sort of control clearly obtains in the case of a battery of ten genes, the homeotic, or Hom, genes, which in the fly embryo determine the identity of the developing segments. This battery of Hom genes establishes a gradient along the antero-posterior axis of the animal that permits the cells to find their way along the axis and to locate their position, which commits them to a given developmental pathway. It is the mutation of one or another of these Hom genes that results in an error in segmental identity and, for example, the formation of a fly with four wings instead of two, or a leg in place of an antenna.

In this fashion, we begin to see "how a fly is built," as my little class-mate used to say. The animal is constructed in the form of repeating metameres, that is, segments or multicellular modules. Cellular dif-ferentiation occurs within these segments along well-defined path-ways. The function of the genes that, in the embryo, are the first to be turned on is to delimit the boundaries of these modules, which at first are all identical, all copies of a modular prototype. The role of the genes causing homeotic mutations, the Hom genes, is to modulate, in each segment, the general rules that will assure the differentiation of the standard segment. Consequently, in each segment these genes di-rect particular modifications that will determine its developmental fate. In fact, each segment, each region is determined by a specific combination of several Hom genes functioning in parallel in every cell in the region.

The existence of mutants reveals another aspect of the mechanism. When a leg replaces an antenna on the head of the second mutant, it is because a gene has been inactivated by the mutation. In a normal fly, this gene functions properly. And it is the functioning of this gene that, among others, prevents a leg from sprouting in the place of an an-tenna. Finally, the mutants offer the opportunity to study the genetic system that underlies the development of the embryo in *Drosophila,* particularly since the fly lives two lives: first as a larva, then as an adult. Many mutations that are not compatible with the life of a fly none-theless permit the formation of a larva, even if modified, even if grossly deformed. The question then becomes how to find a way of selecting the mutations that will modify not the materials for con-structing the fly but the elements that put the materials in place and define the form of the animal.

Homeotic genes work by modulating the activity of target genes that determine the production of growth factors and their receptors. When expressed, these target genes regulate the proliferation and in-teractions of cells. It is, therefore, a hierarchy of genetic elements that directs the development of the fly. Antonio García-Bellido calls the Hom genes and their target genes "selector genes" and "activator genes," respectively, to reflect the role they play in this hierarchy.

We still have only a vague notion of the way in which higher structures in the animal are organized. In particular, we do not know how the size and shape of the body and its organs are determined, that is, which processes regulate cell growth and morphogenesis. Most probably, the requisite information is contained within the individual cells, and the programmed genetic behavior of the cells determines the organism's supracellular organization. In all likelihood, the organization of the body as a whole is the result of local interactions between groups of cells.

Naturally, the structure of the embryo doesn't emerge out of nowhere. The maternal genes orient it and put it in place. The formation of a fly begins in the mother, when she begins to prepare the egg. This egg contains all the building materials proteins, fats, sugars, mitochondria, and the nucleus, in which the instructions required to produce the fly are inscribed. In addition, the egg must contain "something" that organizes all this material, distributes it in space, prepares for the emerging organs, sketches the shape and structure of the future organism, and establishes the cues that will tell the future cells where they have to go and what their role will be—that, in short, gives the matter its form. This "something," this "positional information," turns on the genes in the nucleus that come half from the father and half from the mother. And this interaction, this process of informing, is what makes a fly a fly and an elephant an elephant. This positional information, already present in the egg, must necessarily have been put in place by the mother in the course of constructing the egg, during oogenesis. It must therefore be controlled by the maternal genes, not the embryonic genes. Only after this system is in place can the genes of the embryo be switched on in their turn and go to work.

It is interesting to note that the anteroposterior axis the mother contributes to the embryo is the same as the mother's own antero-posterior axis. In other words, ever since there have been flies and these flies have reproduced, they have faithfully perpetuated the same anteroposterior axis, in the same direction. Obviously, this axis doesn't emerge out of nowhere, either. It comes from another organism from which *Drosophila* descended in the course of evolution.

In literature, the fly is a familiar insect, a symbol of ridicule and annoyance. Less repulsive than a spider, less attractive than a butterfly, it embodies restlessness, derision, and uselessness. Worse, it's a distraction. Montaigne complains that its buzzing "is enough to murder [his mind]," Pascal that it "paralys[es] his reason." And its predilection for garbage, rot, and dung make it a disgusting thing.

Only in science did the fly succeed in becoming a star, not only in genetics, but also in embryology, and once again in a totally unpredictable manner. Embryology remained for a long time a closed discipline—you might almost say an area managed by a cluster of tribes, each devoted to its favorite organism. The embryo of a sea urchin does not resemble that of a frog, a mouse, or a fly. The development of each appears to depend on mechanisms that do not a priori have anything to do with each other. Once the embryonic development of the fly was broadly understood and the genes involved had been identified, the amazing thing was to discover that the same genes—with closely related functions—kept turning up in the most diverse organisms. Common principles seemed to underlie the development of all embryos. The fly thus became a sort of ideal model. Whatever progress we are able to make today in the genetic study of mice or humans, we owe to the fly.

3   The Mouse

....

Over the course of this century, the number of authors listed on biology research papers has increased dramatically. Before World War II, each published paper usually represented the work of a single scientist. After the war, the count was generally two per paper. Since the 1970s, the number has risen steadily, to the point where in recent years it reached several dozen or even a hundred in work on genomes.

In the early days of molecular biology, at mid-century, most research was the product of teams of two, of duos, of pairs. Take, for example, the work of George Beadle and Edward Tatum, Salvador Luria and Max Delbrück, Max Perutz and John Kendrew, James Watson and Francis Crick, Jacques Monod and myself, Matthew Meselson and Franklin Stahl, and others. The papers published by each of these pairs described major steps in the development of molecular biology. Why did pairs play such an important role? What made this era, this field, so favorable to scientific partnerships? Was it the interdisciplinary character of the research? Or the breadth and diversity of the techniques, derived from different fields, employed in the research? Or the complexity of the experiments?

These suggestions strike me as unlikely. It seems to me rather that what really allowed these pairs to exploit their talents and to demonstrate their effectiveness was the theoretical side of their research, as opposed to the experimental side. When a science is in its infancy—when the landscape is open and indistinct—there is plenty of time to dream up theories and construct models. And two are better than one for dreaming up theories and constructing models. Interior monologues are less conducive to this kind of exercise than are the dialogues between two people accustomed to cooperating, discussing, commenting on each other's work, and balancing two different ways of looking at the world; in short, the interaction of partners used either to working together or against each other. To say nothing of the amusing side of research, it's more enjoyable to work in pairs than alone. For with two minds working on a problem, ideas fly thicker and faster. They are bounced from partner to partner. They are grafted onto each other, like branches on a tree. And in the process, illusions are sooner nipped in the bud.

In fact, with pairs, work rapidly takes a different turn. It follows rules unique to the two associates and to the game. Like twins, partners use a special vocabulary. They invent new words. Often, during particularly lively discussions, when the dialogue bounces along like a ping-pong match, the excitement can reach a level where each player responds before the other has finished a sentence. So much so that any outsider present at the scene very quickly loses track of the discussion.

Twice at the Pasteur Institute, I had had the opportunity to participate in such a partnership—first with Elie Wollman in studying lysogeny and sexuality in bacteria, and later with Jacques Monod in analyzing the inducible synthesis of proteins in the bacterium *Escherichia coli*. Lewis Thomas suggested the importance of a piece of research can be measured by the intensity of the surprise it provokes. Well, there was certainly no shortage of surprises at the Pasteur Institute. With Wollman, the so-called erotic induction of the development of a prophage during conjugation; the kitchen blender-induced coitus interruptus of mating bacteria; the circularity of bacterial chromosomes. With Monod, the so-called PaJaMo[1] experiments that I described in *La Statue Intérieure* (The Statue Within) and, especially, a whole battery of mutations. Some of these were totally unexpected, like the dominant negative mutations[2] we identified in the regulatory loops of bacteriophage lambda and the lactose system. Another kind of surprise cropped up in the course of this work: after producing a model that we had trouble taking seriously ourselves, we were astonished to find that it contained a portion of truth; that the world, or a little fragment of the world, was meekly conforming to what we had imagined, at least for the time being!

For several years, Jacques Monod and I spent hours every day in his office, half the time arguing, half the time drawing schemes on the blackboard. Gradually, thanks to the combination of work on lysogeny and on the lactose system, and by dint of many discussions and a lot of drawing on the blackboard, we came up with the so-called operon model. This model summarized what we knew of the mechanisms of protein synthesis. It advanced the idea of a regulatory unit of gene expression, a loop consisting of a regulatory protein that recognized a sequence of DNA that controlled the expression of adjacent genes.

We proposed this model to explain the regulation of gene expression in bacteria. But our hope was to find, in higher organisms, anal-

1. PaJaMo (for Pardee, Jacob, Monod) refers to work on the regulation of protein synthesis that proved that inhibition (or repression) was the essential mechanism. [Trans.]
2. Dominant negative mutants are mutants that have deleted the region that can initiate transcription of the genes for phage formation. Thus, removal of the repressor does not yield any phage. [Trans.]

ogous regulatory units that functioned according to similar, though obviously more complex, principles—in particular in the phenomena underlying embryonic development and cellular differentiation. These regulatory units could be considered to work like elements of electronic circuits. They could be combined at will, and linked to activate or repress gene activity in a number of ways in response to diverse regulatory functions.

Upon publication, the operon model was an immediate hit. It may in fact have been too big a hit. For it was applied in every way imaginable, even in areas that had nothing to do with it. Different models were proposed to explain regulation of gene expression in bacteria by various mechanisms, some radically different from ours. These were rapidly debunked by biochemists. Not, unfortunately, by those of the Pasteur Institute, who never approached the problem in the right spirit or with the appropriate resources, but by those at Harvard— Walter Gilbert and Benno Müller-Hill, who succeeded in isolating the lactose system repressor, and Mark Ptashne, who purified that of the lambda phage. Through these efforts, the molecular study of regulation became possible. The elements discovered did indeed make up a regulatory unit whose properties corresponded, with only a few minor modifications, to those predicted by our model.

What interested me from that point on was to find out whether the principles uncovered in the regulation of gene expression in bacteria also functioned in complex organisms, and especially in embryonic development. If they did, I wanted to understand the origin of the necessary additional complexity. This was a particularly difficult problem to tackle, because at the time it appeared that applying the methods used to study bacteria to complex organisms was out of the question. With bacteria, locating and analyzing the elements of regulation had depended almost entirely on the genetic analysis of the cell—analysis that at the time was inapplicable in so-called eukaryotic organisms. The one exception was yeast, which was perfectly suited to this kind of study. Apart from certain phenomena related to sporulation and sexuality, however, studies of yeast had almost nothing to do with embryonic development and cellular differentiation.

If I really wanted to study differentiation, I was going to have to choose one of two paths. I could either stay with the study of unicellular organisms—bacteria or yeast—and explore new lines of research, or, instead, I could go for something bigger and more complex, which meant finding a different organism. Difficult decision. Few scientists, and certainly few biologists, change once they've embarked on one area of research with a particular organism. They while away their existence more or less successful in their inventions. It's a rare person who can, like André Lwoff, repeatedly switch both area and organism, and still bring a fresh approach to an old problem every time.

Obviously, I had to discuss this question of organism and change with Jacques Monod. Since the work on the operon and our inability to isolate the repressor at the Pasteur Institute, our interests had gradually diverged. Jacques had devoted himself to studying the characteristics of certain regulatory proteins whose properties, called "allosteric,"[3] had been described in the work of Jean-Pierre Changeux. I had taken part in this work at the beginning, when the evidence showed that the proteins involved in the regulatory loops—repressors or activators—presented allosteric properties. In particular, it was one of the only ways of explaining the characteristics of certain of my favorite mutants, for instance, the dominant negative mutants. But developing models and testing them depended above all on detailed kinetic studies of the chemical reactions displayed by these proteins. And these kinetic studies just didn't turn me on.

So I turned my attention to the study of cellular division and DNA

3. "Allosteric" refers to a class of enzymes that undergo a change in shape that modifies their function. [Trans.]

replication in *E. coli*. In 1962, the Brenner family came to spend their vacation with the Jacob family by the beach at La Tranche-sur-Mer (which Sydney called "The Slice") in Vendée. While the children played on the beach, Sydney and I talked and sketched in the sand. We finally produced a model called the "replicon," which linked cellular division and DNA replication. The idea of the model was to attach the DNA to the bacterial membrane at a site that controlled replication. Using this model, we tried, with François Cuzin, to isolate and analyze *E. coli* mutants in which division and/or replication had been disrupted.

But the question that really mattered to me was, should I change the organism I worked on and my area of research? Although we didn't work as closely together as before, Jacques and I still talked to each other very often. I had tried several times to steer the discussion to the subject of a new direction, but Jacques wasn't too interested. He was still very attached to the study of microorganisms, which he considered—and rightly so—to be the organism of choice for studying many unresolved problems.

I didn't find his arguments very convincing, though, for several more or less valid reasons. First of all, I had no desire to spend my life doing the same experiments. Despite Alfred Hershey's quip that, for the biologist, happiness is getting a complicated experiment to work and then repeating it in exactly the same way day after day, I wanted change. For some fifteen years now I had been monotonously watching matings of carefully chosen pairs of bacteria. This sort of exercise had given me a lot of satisfaction. But I felt I had exhausted its delights. (Not that I was opposed to becoming a sort of sexual guru; I just didn't want to specialize in bacterial sex.) Second, bacteria were beginning to seem a little tame, a little reserved, to me. I wanted to work with something you could see, that had hormones, passions, a soul. I wanted to work with animals visible to the naked eye that I could distinguish individually, even name. Animals capable of looking me in the eye. Though even that wasn't the real reason. The real reason was more serious, more biological, more professional. Some of the phenomena associated with microorganisms were not unlike cell differentiation—in baccili, sporulation; in yeast, especially, which are very close to the cells of higher organisms, certain aspects of sporulation and sexuality. But that was cheating a bit. To study the development of embryos, you had to work with embryos. To study cellular differentiation, you had to work with organisms with differentiated cells: muscle, nerve, skin, kidney, and so on. There was just no way around it. You had to do it. I decided to change.

Once I had made this decision (around 1967), two questions came up. What organism would I choose? and, How would I move from one type of research to the other? In biology, the choice of which organism to study is extremely important. First, because the very nature of an animal, that is, its structure and physiology, limits the research possibilities to certain types of experiments. Second, because over time you become a sort of captive of what you have done and what you know. Equipment and other paraphernalia—mutants, enzymes, purified products, all of great value—pile up. Commitment to a particular area of research, with a particular organism, also represents an investment of time and work that compounds daily. If, once you've started down a certain path, you have a change of heart, it can be difficult to shift direction.

How would I choose among the favored organisms of embryologists—sea urchin, frog, fly, mouse, and so on? Each was well suited to one particular type of experimentation, but not at all, or only somewhat, to others. One day I took a piece of paper and noted down all the properties I thought would be desirable in an animal to satisfy the needs of the type of research I wanted to do. It had to be easy to raise, quick to reproduce, simple to analyze genetically or biochemically. I wanted to be able to produce cell cultures from this organism and study its physiology and behavior. And so on. It was clear that the ideal animal did not exist. Only something like a hybrid of frog, sea urchin, and fly would meet such requirements. I had to back down and compromise.

In the course of a discussion in the lab, a researcher suggested we consider planarians (flatworms). The advantage of these worms is their ability to regenerate. When you cut a planarian in two, each half reconstitutes an entire organism, and you obtain two planarians, each with a full complement of tissues and organs. Thus you can study both the formation of the animal and the differentiation of its cells. We learned that the big planarian expert in Europe was an Italian, a professor of zoology at a university in northern Italy. Three of us went to visit him and see his collection of planarians. He was a dry little man, but pleasant, with black hair and eyes. He told us straight off how honored he was that molecular biologists were interested in his humble organisms. To our first question, how long does reproduction take in planarians? he answered frankly, "About three months." Seeing our crestfallen faces, he added: "Maybe two and a half months." And when this concession didn't seem to cheer us up, he ventured with a marvelous lilting accent: "Maybe ten weeks. I can't do better than that." Discouraged, we prepared to return to Paris.

The zoology professor then asked if I would give a seminar for the students. We settled on the subject ("DNA Replication and Cellular Division in Bacteria") and the time: the next day at eleven o'clock. The next day I arrived at the university at ten-thirty. But it was 1968, and I found my zoology professor in a state of total confusion. "The students have been occupying the auditorium since last night. You can't give your seminar." He sent one of his assistants to negotiate with the students anyway. Finally, at five to eleven, the students agreed that I could give my seminar. At precisely eleven, the zoology professor and I entered an auditorium filled to overflowing, feeling like condemned men tossed to the lions. Scarcely had the professor gotten up to introduce me when the students began to catcall. Eventually he sat back down. The racket stopped. I stood up and told my little story of *E. coli* mutants incapable of reproducing their DNA to a silent crowd.

When I had finished, everyone applauded. Then the questions started to fly. "Have you isolated the mutant enzymes?" "Do you think students can start a revolution?" "How many different mutations have you isolated?" "What do you think of the anatomy professor's attitude toward the students?" In the end we agreed to group the questions: first the scientific ones, then the political ones. So, to begin with, I discussed a few points from my lecture. When that was finished, all the professors who had sat through the session, including my host, the zoology professor, got up and left. I was alone with the lions. A series of questions followed, which I answered somehow or other. Some concerned the general situation, the evolution of our society, the need for political change, and so on. Others specifically concerned the situation at that university, about which I knew nothing but was rapidly getting a feel for. The students were particularly angry at the anatomy professor. They thought he had treated them badly. It all went very well, with a good dose of humor. After three hours of discussion, we stopped. As I left, the students were chanting: "This is only the start! Let's not give up the fight!" Once outside, I was amazed to find that the entire faculty were waiting to take me to a lunch nobody had told me about. All the professors hurried toward me. How did it go? What questions did they ask? What do you want? And that's how I learned, to my surprise, that the professors of that university had not talked to their students for three months. That was the end of the planarian episode.

Around the same time of year, I was in New York with Seymour Benzer and Sydney Brenner to attend a conference. One evening, before eating, we went to the movies to see *The Mouse That Roared*. The film is about a little European country governed by a grand duchess. During a Cabinet meeting to discuss the country's catastrophic financial situation, a Cabinet minister remarks that the only countries whose financial situations are flourishing are those that were defeated in the Second World War—Germany and Japan. So the prime minister proposes that war be declared on the United States. The proposal is enthusiastically approved. An expeditionary force of six men led by a sergeant is then sent to fight the United States. It so happens that the day the expeditionary force lands in New York is the Fourth of July. The city is deserted. Not a soul in the street. Not a single car. The troops disembark and take over the city without firing a shot. The comic effect is reinforced by the device of having the English actor Peter Sellers play all the leading roles, from the grand duchess to the prime minister to the sergeant.

After the movie, dinner. And there we started a good discussion. Seymour and Sydney had already taken the plunge. Both had abandoned phage and bacteria several months earlier. Both had decided to study the connectivity of the nervous system, using cleverly selected mutants—for Seymour, *Drosophila,* for Sydney, a little nematode called *Caenorhabditis elegans.* Each was boasting about the advantages of the organism he chose, which was obviously superior to the other's. The fly already had a distinguished scientific past, having long been the geneticists' organism of choice. Quick to reproduce, flies were easy to raise in the laboratory; a simple bottle could accommodate hundreds. Moreover, a collection of highly varied fly mutants was available that included some mutants whose embryonic development had been disrupted. It was thus an ideal organism for studying the genetics of development. It did, however, have some drawbacks. Its physiology was difficult to study, and its cells nearly impossible to cultivate.

Sydney's little nematode was a different story. Rarely used at the time as an experimental subject, its principal virtue was its ability to reproduce rapidly on gel plates seeded with *E. coli.* The nematode was a little roundworm, a hermaphrodite, half a centimeter long. One individual could produce a hundred descendants in three or four days.

The animal itself was simple to breed and easy to analyze genetically. Sydney had already isolated a series of mutants. An astonishing property of this little worm was that it contained only nine hundred and eighty-two cells, the fate of each being fixed with precision. But though you could characterize it biochemically, you couldn't really examine its physiology. And as with the fly, the techniques for cell culture using nematodes were primitive. In the end, one system was as good as the other. One had the advantage of familiarity, the other the advantage of speed.

To move from one area of research to another, it was necessary to adapt researchers, materials, and my laboratory to the new regime. I thought that it might be possible to take a first step in this direction and to become familiar with mammalian cells by cultivating mouse cells. After all, André Lwoff had made this kind of shift effortlessly when he gave up phage for poliomyelitis virus as an object of study. And Boris Ephrussi in France and Henry Harris in England were attempting to study cellular differentiation in cell cultures. They "fused" different types of differentiated cells and looked at the expression of different characters in their products. There again, this kind of experiment seemed to me to be cheating a little to avoid having to deal with embryos. But it was a beginning.

It didn't cost much to convert the bacteriology laboratory into a cell culture laboratory. David Schubert, an American, came to spend a two-year sabbatical and brought some mouse neuroblastoma cells with him. Then Hedwig Jakob, a collaborator of Ephrussi's at Gif-sur-Yvette, decided to move to Paris. She was an expert "cultivator" of cells, comfortable with all the techniques. Over several months, we put into culture several types of cells, especially mouse lymphocytes. We were trying to have them reverse certain aspects of differentiation by selecting cells capable of synthesizing an enzyme that usually disappeared after lymphocyte differentiation. A fruitless exercise. We were vaguely considering fusing an antibody-synthesizing lymphocyte with a non-antibody-synthesizing one when the news came that Cesar Milstein and Georges Köhler had discovered monoclonal antibodies!

These activities were merely a diversion before we could commit ourselves to embryos. The small laboratory I occupied was difficult to remodel and could accommodate no more than a tiny organism. Fly or nematode? For several months I had been playing around with the strain of nematodes that Sydney had given me. Françoise de Vitry, Hedwig Jakob, and I produced several mutants from them. We tried our hand at some genetics and biochemistry using them. But those little worms didn't excite me at all. At the end of six months, I gave a lecture at the Pasteur Institute titled "What You Can't Do with a Nematode."

Drosophila was more appealing to me on account of the enormous variety of mutants that had accumulated over fifty years of laboratory research, some of which clearly appeared to disturb the formation of the animal, and consequently its embryonic development. And also on account of some experiments done in Switzerland by Ernst Hadorn, who had used the abdomen from an adult fly as a test tube in which to cultivate embryonic cells.

But on reflection, was it sensible to work on the fly at the Pasteur Institute? It seemed to me unthinkable to leave Pasteur for another center of research. For a lot of reasons, I wanted to stay on where I had begun, twenty-five years earlier. I was particularly conscious of what I felt was my debt to the Institute and to the Pasteurians who had welcomed me so warmly and had helped me get off the ground. But working at the Pasteur, it seemed to me, implied certain constraints. Intended by its founder to study and to fight infectious diseases, research at the Pasteur Institute had concentrated on the study of bacteria, viruses, and defense mechanisms for seventy years. If molecular biology had managed to develop there, it was precisely because in its infancy the experimentation built on work with bacteria and viruses.

It was certainly pointless to work directly within the very disciplines that mobilized, so to speak, the main battalions at the Pasteur. But it also seemed futile to introduce a new organism, given the logistics of the support it required.

On the other hand, work was already going on at the Pasteur, though to my mind not enough of it, with an organism central to the Institute's main missions: the mouse. This little animal was well suited to immunological study, but the immunologists used rabbits. You could infect the mouse with certain bacteria or pathogenic viruses, but the bacteriologists mostly studied guinea pigs. Finally, the mouse was the perfect animal for studying certain cancers and transplantation. Besides, genetic analysis of the mouse had been under way since the beginning of the century. It reproduced more rapidly than other mammals. Many mutations had already been identified. Several laboratories had undertaken the study of the mouse embryo. Clearly, analysis of the development of an embryo implanted in the maternal uterus was much more complex and presented many more difficulties than in a fly, a nematode, or even a frog. But the mouse had two virtues that at the time were decisive for me. First, it was the smallest mammal, the laboratory organism similar to man with which studies in genetics, physiology, and pathology would serve as models for humans. Second, and more important, it met the needs of most of the research being carried out at the Pasteur Institute. It was thus perfectly logical to study the genetics and developmental biology of the mouse there. That wasn't the case for the other organisms, such as the fly, the nematode, or the frog.

The more I thought about this business of mice, the more obvious it was to me that we would have to adopt the same approach that had worked so effectively with bacteria. In particular, we would have to get researchers from different disciplines working together on the same organism, that is, the mouse. At the time, government administrations were broadly sketching out what they foresaw for the future of France, in very diverse areas, in the shape of "five-year plans." At the end of the 1960s, a plan was thus being prepared for the years 1971–1975. Since I was among the consulting scientists, I suggested that we build a mouse institute where geneticists, physiologists, biochemists, pathologists, virologists, oncologists, and so on could work side by side on the same organism. The suggestion was not well received, in the first place by the science administrators. They asked me straight off whether I wanted to head the Institute, a job which didn't interest me at all. If I had accepted, they would perhaps have built the institute. But in their opinion, such a construction project was conceivable only for a person, not for a research program! The response from my colleagues was just as bad. They had three principal objections: (1) He wants to build such an institute to be able to direct all of French biology. He's a dictator. (2) It's a stupid idea. People should be left to work alone as they like. (3) Why mice? Why not a sea urchin institute or a frog institute? If I had wanted to direct such an institute and had fought hard enough for the project, I would probably have succeeded. But I found the criticisms and comments so inept that I didn't even try. I may have been wrong. At the time, building such an institute would very likely have given French research teams a good lead.

To work with mice, I would need room. There was no question of finding any on the old premises of the Pasteur Institute, which had seen the beginnings of molecular biology. To develop the new biology, a new building was going to be constructed. It wouldn't be ready until the beginning of the 1970s. While I waited, I continued my training in mammalian cell culture with Hedwig Jakob.

My enthusiasm for studying mice grew when I learned about the existence in this organism of a tumor that had unusual properties. It was called a teratocarcinoma and was found in the testicles of certain inbred lines of mice. These tumors could be transplanted serially from mouse to mouse. They each contained widely differentiated cell types—muscle, nerve, gland, heart, and so on—as well as an undifferentiated type of cell similar to the cells of early embryos. Several laboratories, Boris Ephrussi's among them, had succeeded in isolating the embryonic cell type and keeping the cells in test tubes. Grown in culture for several months, then reinjected into mice, the embryonic cells retained their ability to induce tumors that contained a wide range of cell types. Moreover, certain indications led us to believe that the cells could even differentiate in the test tubes. These tumors were extraordinarily interesting to me. First, in all likelihood they would allow study of differentiation of cells simultaneously in vitro and in vivo—in an embryo. Second, they would make it possible to examine a supposition I had considered for a long time: that there was a close relationship between embryos and cancers. It was the only explanation I could find for the fact that often the molecules and proteins appearing in cancers did not show up in adult organisms. Since these molecules couldn't have fallen out of the sky just like that, they had to come from genes expressed at certain stages in the development of the embryo, and not in the adult. Oncogenesis most likely had the effect of disrupting certain regulatory systems at work in the development of the embryo. The teratoma, or rather the teratocarcinoma, of the mouse would perhaps allow us to work out this relationship.

So it would be the mouse—cells and embryos. At the end of the 1960s, the decisions had been made. I still had to wait for the molecular biology building at the Pasteur Institute to be completed. In addition, two conditions had to be met to get the plan under way. First, the future building would have to be optimally adapted for work on mice. The building was destined for the molecular biology teams, which meant the teams in the laboratories of André Lwoff, Jacques Monod, and me. Until then, all these people had worked with bacteria. Except for a group that had begun to study polyoma virus and its reproduction in animal cells, no one had yet talked of moving from bacteria to a more complex organism. Yet, so far as we could see, the research on mice was going to require a certain critical mass. Toward this end I began to generate a little propaganda in favor of mice, with the ulterior motive of persuading enough other groups to create a sort of miniature mouse institute in the new building.

I talked about these plans several times with Jacques Monod. He was not yet director of the Pasteur Institute, but it was clear to everyone that he was going to be. In fact, his nomination had been delayed several months for a surprising reason: the opposition of President Pompidou. The night the Nobel[4] Prize was awarded, Jacques had given an interview to Jean Daniel for *Le Nouvel Observateur.* In it he criticized the scientific policy of the government headed by Georges Pompidou and stated that Pompidou was uninterested in science. Pompidou, who like all presidents of the Republic, had the memory of an elephant and the spitefulness of a rhinoceros, had not forgiven Jacques for the remark. He remained opposed to his nomination as director for several months. It was a surprising reaction, first, because, according to the statutes, the president of the Republic had nothing to do with the nomination of the director of the Institute. It was also surprising because Pompidou's memory was selective—he had forgotten something de Gaulle once said. During a Cabinet meeting, a minister who suggested making some changes to the Collège de France had provoked the following remark from the General: "There are three things in France that are inviolable: the Collège de France, the Pasteur Institute, and the Eiffel Tower."

Monod was not very enthusiastic about the prospect of my switching from bacteria to mice. He thought, with good reason, that there was still a lot of work to be done to gain a better understanding of the

4. François Jacob, André Lwoff, and Jacques Monod shared the 1965 Nobel Prize for physiology or medicine for their work in genetics, especially the operon theory of gene regulation. [Trans.]

functions and genetics of the bacterial cell. He thought it was a shame
to rudely break up teams that had collaborated very effectively. And
he'd add, always to the point: "You are going to weaken the old groups
and find yourself isolated in a world about which you know only a little—or perhaps nothing at all." Finally, after several discussions, he
was persuaded that it was not totally unreasonable to take on regulatory systems in higher organisms, and in particular in mice.

One important issue remained that would be resolved once Jacques
became director of the Pasteur Institute: the animal room. At the
time, the Pasteur Institute was very weak in this respect: too little
space, bad equipment. A big animal room had been foreseen in the
basement of the biology building that was under construction. The
question was, what animals were we going to put there? To my mind,
the research to be carried out on the mouse was going to require the
maintenance of several inbred lines of mice, isolation and upkeep of
mutants, mapping of mutations on chromosomes by crossing many
animals, and the steady production of embryos. In short, it would be
necessary to maintain a population of several tens of thousands of animals. At the time, the principal users of animals at the Institute were
the immunologists, who preferred to work with rabbits. These immunologists, and particularly their leader, Jacques Oudin, who was
doing superb work on the immunological properties and structure of
antibodies, naturally wanted to fill the animal room of the molecular
biology lab with their rabbits. We were at loggerheads. Rabbits! Mice!
How many of each? Jacques Oudin was determined to keep all the
Institute's rabbits there. I wanted to keep an army of mice there. We
had many discussions—and a few angry words—about it.

Very fortunately for us, Oudin's work convinced him that genes
were directly involved in the synthesis of antibodies. Because there
were at the time no genetic studies of rabbits, he gradually conceded
that mice might be better than rabbits for work on genetics. And in
the end, Monod, by then director of the Institute, decided in favor of

mice. A young veterinarian from the CEA,[5] Jean-Louis Guénet, was recruited to run the animal room. Being both very energetic and very interested in the genetics of the mouse, Guénet turned his animal room into a remarkable tool.

One final difficulty remained. Everyone in our group, which had specialized in the genetic analysis of bacteria, agreed to make the leap to studying mouse embryos. But no one had any experience either in embryology or with mice. We would all need to do an apprenticeship. So Charles Babinet and Hubert Condamine, two researchers who had been working with me for a long time, left for postdoctoral stays— one went to England, the other to the United States—in laboratories specializing in research on the development of the mouse embryo. And Robert Fauve, a former student of René Dubos in New York, came over to Paris to share his expertise in the culture and pathophysiology of mice with us.

In this fashion, little by little, a new team formed. And when the new building was ready in January 1972, all the members of the group applied themselves single-mindedly to the study of teratocarcinoma and mouse embryos. A new life was beginning. It was taking shape at the very moment that genetic engineering was transforming the study of higher organisms. This was to turn our idea of the living world upside down.

5. The French Atomic Energy Authority. [Trans.]

4   The Erector Set

....

Daedalus occupied a special place among the heroes of Greek my-
thology. At once a blacksmith, architect, sculptor, and engineer, he
knew how to work iron as well as wood. He was descended from the
royal house of Athens and had been given his talent by Athena herself.
He claimed a number of inventions. But many people in Athens contested his claims. In his workshop Daedalus had an apprentice, Talos,
who was also his nephew. Although he was only twelve, he already
surpassed his master in inventiveness and ingenuity. One day, Talos
found the jaw of a serpent and realized that he could use the same principle for a device to cut a staff. He fashioned a sort of blade with iron
teeth and thereby invented the saw. This invention, along with others,
such as the potter's wheel and the compass for tracing circles, guar-
anteed Talos an exceptional reputation despite his youth. But Daedalus claimed to have forged the first saw himself. Jealousy soon con-
sumed him. One day, under the pretext of showing his nephew some
architectural detail, he took Talos up to the roof of a temple and
pushed him off. Accused of murder, Daedalus was forced to flee. He
took refuge in Crete.

The people of Knossos knew only about Daedalus's brilliance and his extraordinary abilities as an artisan. So King Minos welcomed him with open arms and gave him the means to develop his talents. Curiously, Daedalus never used his skill in the service of any personal ideology or ambition. Rather, he helped others carry out their harebrained schemes. That's how it happened that, one day, King Minos's wife, Pasiphaë, came to ask Daedalus for help. She had fallen madly in love with a magnificent white bull sent by Poseidon as an act of revenge on Minos because he had violated an oath. Pasiphaë begged Daedalus to give her the means to satisfy her passion. He immediately built her a hollow wooden cow, and covered it with the skin of a real cow. Then he explained to Pasiphaë how to use it: she should open the door concealed in the cow's back and creep into the interior, sliding her legs into the hind legs of the cow. After which Daedalus discreetly withdrew. Pasiphaë followed his instructions. Impatient with lust, her darling bull mounted the cow at once. A few months later, consistent with the genetics of the time, Pasiphaë gave birth to the Minotaur, a creature with the body of a human and the head of a bull, which would eat only human flesh.

Next, Daedalus had to yield to the demands of Minos. Enraged at his wife's infidelity, Minos made Daedalus build a prison whose halls connected in a network so complex that, once inside, no one could ever find a way out. The Minotaur was confined in this labyrinth. Every year Daedalus, who was charged with feeding him, brought him seven young boys and seven young girls supplied by the city of Athens.

Determined to kill the Minotaur, and thereby free his city of the terrible human tribute, Theseus one day surreptitiously joined the young men of Athens who were to be delivered to the monster. Fortunately for Theseus, Ariadne (the daughter of Minos and Pasiphaë) had seen him and had fallen in love with him. She approached Daedalus and entreated him to help Theseus find his way out of the labyrinth. In the blink of an eye the architect flew into action. He gave

Ariadne a ball of string and told her how to use it. She was to stay by
the door and hold one end of the string while Theseus held the other
as he advanced. Theseus killed the Minotaur and, guided by the string,
was able to exit the labyrinth easily. Furious at Daedalus, Minos
locked him up in the labyrinth with his son, Icarus. We all know the
end of the story: in order to escape, Daedalus constructed wings out
of bird feathers that he and his son affixed to their shoulders with wax.
Despite warnings to fly low, Icarus became drunk with pride,
strength, and speed, and intoxicated with air and sun. He flew too
high, fell, and drowned before the eyes of his father, this time helpless.

Daedalus was never a central figure in Greek myth. Yet his role was
often decisive. He personified the skills it takes to control the world.
He had an answer to all practical questions. Once, a friend asked him
to solve a difficult problem: how to pass a thread through a snail shell.
Daedalus thought of a solution at once. He attached a fine thread to
an ant. Then, at the top of the snail shell, he pierced a hole and daubed
its edges with honey. The ant immediately took off at full tilt along the
spiral of the shell, gorging itself on honey as it exited the hole.

Daedalus was a marvelous technician, but he was never more than
a technician who used his skill to serve his masters. He never sought
power himself. He never tried to satisfy any personal ambition or passion. Unlike the heroes he was induced to serve, who would stop at
nothing in pursuing their goals, Daedalus himself always respected the
boundaries of established order and law. He never let himself get carried away by what the Greeks called "hubris."

Hubris is the excessive pride that brings about disorder, the fre-
netic ardor that begets quarrels and confusion. Hubris, says Jean-
Pierre Vernant,[1] leads men to provoke the gods, to place themselves
above human laws. It was hubris, for example, that moved Prome-

1. Jean-Pierre Vernant, *Myth and Thought among the Greeks* (Paris: Maspero,
1965), p. 11.

theus to defy Zeus. Prometheus sought power through knowledge. To attain his goal, any stratagem sufficed. Not so for Daedalus. He fancied himself an engineer—the best of engineers. To keep that distinction, he didn't hesitate to kill a perceived rival. But Daedalus didn't kill Talos out of hubris. Talos's murder was an act of petty jealousy by a man who wanted to steal someone else's discovery, a cheat who attacked from behind out of cowardice. No defiance of the gods motivated this act, no violation of divine law, no attempt to shake the established order, rules, or values.

Yet though Daedalus never lost his head, though he respected the moral and religious dictates by which the gods ruled people's lives, he placed himself entirely at the disposal of others. His skill permitted his masters to abandon themselves to *their* hubris. Through Daedalus and his craft, Pasiphaë, Minos, Theseus—even Icarus—were able to give themselves over to their reckless ambitions and to pursue their passions to the limit. In this sense, Daedalus symbolizes an evil of our age: the high-flying technician who uses his talent to serve any ideology without worrying about its content or worth. Daedalus was the epitome of "science without conscience."

To the Greeks hubris was evil. A world given over to hubris was a world turned topsy-turvy and vulnerable to disorder. A world where force alone dictated law and where people were abandoned to pain and wretchedness. This word "hubris," this old Greek word, is perhaps, according to Lewis Thomas,[2] the word that best explains the fear and apprehension that for some years have been evident in the public's opinion of science and scientists. Hubris refers not only to what many see as an unbearable meddling by scientists. It encompasses all the products of science and technology that threaten the future of the planet and its inhabitants at this century's end: atomic energy (both bombs and nuclear power) and industrial excesses (including pollu-

2. Lewis Thomas, *The Medusa and the Snail* (New York: Viking, 1979), p. 65.

tion, the greenhouse effect, and dangers arising from undersea oil exploration)—in short, everything we hold responsible for the deterioration of our world.

For many years public disapproval focused on physics and its allied technologies. Biology, considered the auxiliary engine, if not the motor, of medicine was spared. Biology contributed to humankind's perpetual efforts to conquer illness, pain, and suffering. In recent years, however, biology has begun to receive its share of condemnation. Public opinion falls hard on everything it holds responsible for the deterioration of the world and everything it judges to stem from the foolish hubris of scientists. In the same shopping basket of opprobrium we find a jumble of fears, some realistic, some fantastic: chemical control of human behavior, various organ transplants, global overpopulation, cloning of humans (production of thousands of copies of a single person from a snippet of skin), test-tube babies, and, especially, genetic manipulation and the creation of monsters. Even the thought that we can take genes from one organism and insert them into another is intolerable to many. The notion of recombinant DNA is tied to the mysterious and the supernatural. It rekindles the terror associated with the hidden meanings of monsters, the revulsion engendered by the notion of two beings merged in defiance of nature.

Some twenty years ago, newspaper headlines announced the latest miracle of science to a world seized simultaneously with admiration and alarm: the production of a baby in a laboratory. A group of scientists in England had succeeded in creating a baby whose conception, nine months previously, had taken place not by traditional means but in a test tube. This event, the international press was quick to promise, was going to revolutionize not only biology and medicine but all of society. For millennia, people had sought pleasure without conception. At last it would be possible to have babies without pleasure!

The media treated the news as an unprecedented exploit in life sciences. Yet its novelty consisted solely in transferring to humans the results of experiments carried out fifteen years earlier with mice, the experimental animal closest to man. Using mice, we had learned to collect sperm and eggs and to combine them in test tubes in order to fuse one with the other; to implant the embryos thus created in surrogate mouse mothers; to freeze the embryos; to thaw them and reimplant them years later, producing a hodgepodge of generations. In short, we had achieved total mastery over initiating embryonic development.

What distinguishes basic from applied research is that, in the latter, we know what we will find, whereas in the former we have absolutely no idea. If applied research endeavors to fine-tune the details, to bring the execution of a given plan to conclusion, basic research aims to reach a fundamental understanding. In the case of the test-tube baby, the essential problem was to establish, in the case of a woman, the proper conditions for transferring knowledge acquired from research with mice. The protocols for collecting eggs and the dose of hormones needed to obtain the desired result had to be figured out. Not only did we know the objective, but we could precisely predict the developments, side effects, and possible applications: freezing embryos and reimplanting them in surrogate mothers; producing children from the sperm of deceased donors; actualizing the desire for children for women or couples who could not otherwise have any; trafficking in embryos; mixing of generations so that a great-nephew would be able to impregnate his great-great-aunt, and so on. In the early 1970s, I attended a conference on reproduction in mice. In the evening, after dinner, the participants speculated, in free-ranging discussions, about applying to humans the in vitro fertility techniques already mastered in mice. They also imagined the situations and fantasies that in vitro fertilization in humans was sure to provoke.

It was an uneven chorus of cheers that greeted the first test-tube

baby. Yet, all things considered, it really entailed only a minor mod-
ification of the usual process, a tiny variation in the first of a series of
reactions that happen by the hundreds of thousands in the development of an embryo. The only change was in the place in which sperm
and egg usually meet up—a plastic container instead of the Fallopian
tube. But, of course, there's no stopping progress. Already, talk runs
to extending the test-tube stage to the stages following fertilization.
It is even predicted that, as in Aldous Huxley's *Brave New World*, the
entire development of the embryo—the nine months of fetal life—
will take place in increasingly refined culture media. When that day
comes, it will spark a new chorus of outrage and admiration. Once
again, it will be in all the headlines. Some scientists will quit their jobs
to show that things really have gone too far, that this research that is
leading humankind to ruin must be stopped. Ethics committees will
create specialized subcommittees. Parliament will discuss, as a matter
of lesser or greater urgency, the need to institute a series of new laws.

But what is so extraordinary, so fantastic about the birth of a child
has nothing to do with the nature of the container in which the first
step took place. Nor would it be possible to carry out the entire process in a test tube. What is incredible is the process itself. The meeting
of a sperm and an egg unleashes a huge series of reactions—hundreds
of thousands, one after the other—overlapping and intersecting in a
network of astonishing complexity. All of this happens in the course
of gestation, whatever the conditions of conception, to produce a human baby—not a baby duck, giraffe, or butterfly. Incredibly, once fertilization is over, the first cell—the fertilized egg—begins to divide.
It becomes two cells. Then four. Then eight. Then a little cluster of
cells. This cluster attaches itself to the lining of the uterus, grows, develops, and, a few months later, has formed a baby with (in more than
ninety-five percent of cases) everything it needs to live, travel the
world, and even think. That's the true miracle. It's the most amazing

phenomenon in the world. So amazing that the whole world should marvel at it. People should spend their time wondering about the mechanisms that underlie such a marvel.

Well, aside from the rare expert, no one *is* interested in this extraordinary phenomenon. Nobody talks about it. Certainly not the press. We are so accustomed to it, we are so used to seeing a baby appear nine months after lovemaking, that we ask hardly any questions about what happens between the two big events. It's a fact of nature. That's the way things are. Yet for a long time this process was a total mystery. Just until very recently, we barely had any idea about the forces and mechanisms at play. Only in the past few years have certain aspects of this fundamental mystery come to light. This illumination was the result of a very recent, major overhaul of our way of looking at the structure and function of living beings; of a profound change in our account of the living world.

Another story, no less edifying: cancer. In the 1960s, jealous of the prestige President Kennedy had acquired in pushing to land a man on the moon, President Nixon targeted another dream: a cure for cancer. He declared a war on cancer, and resolved that, by applying the necessary resources, cancer could be beaten in five years. We know how the story ends: a lot of money was spent in vain. To get to the moon, all that was needed was means, organization, and tenacity. But none of the fundamental mechanisms of cancer that coordinate cellular division and differentiation were known. The effort mounted at the time could only fail. Then, in the early 1980s, the scientific landscape began to change. The kind of work that up to then had represented the cutting edge of research on cancer became obsolete overnight. Now the field that the most gifted students had so carefully avoided was pursued by swarms of the most talented. Cancer research suddenly became one of the most exciting and most promising aspects of biology.

What caused this change? Surely not any administrative decision.

Nor a massive injection of money à la Nixon. What happened was simply something that happens from time to time in basic research: an
astonishing series of surprises that clearly could never have come out
of any design or plan. These surprises were the result of genetic engineering applied to questions of basic biology. Genetic engineering,
itself the product of a major surprise at the end of the 1960s, allowed
a gene to be isolated from one organism and to be inserted into
another. It allowed us to obtain certain genes from cancer cells that,
injected into normal cells, made them cancerous. These, then,
were "cancer genes," or oncogenes. Other genes, called anti-
oncogenes, were later found to suppress the effect of oncogenes. In
short, a whole genetic arsenal was closely associated with the battery
of genes that, as might be expected, regulate cellular division in connection with differentiation. Oncogenes cause cells to proliferate.
Anti-oncogenes prevent them from proliferating. Cancers occur because of too many oncogenes or too few anti-oncogenes. For the first
time, research on cancer became a respectable area. For the first time,
we could begin to discern the mechanisms underlying a cell's malig-
nancy. For the first time, we could entertain the possibility, still very
distant, of manipulating these mechanisms. All these advances were
the result of a new way of seeing and studying the cell, of a change in
our understanding of the living world.

The role of science is precisely to describe the world—living beings and nonliving things—within certain constraints: to probe as
deeply as possible beneath the surface of things and their appearance;
to dispell, insofar as possible, the illusions that nature imposes on our
senses and our brains. Our senses and our brains are the products of
evolution. They are adjusted to the life that each organism lives. For
example, evolution has furnished humans the means that permit us to
live on this earth, to perceive in the world around us the objects of
everyday reality, to deal with a memorable past and an imaginable future. As soon as we leave the world that we can apprehend directly,

as soon as we wander from the world of mundane objects, our brains cannot follow. This is what physicists have observed over the course of this century, in trying to analyze the infinitely small and the infinitely large.

You might say that the scientist lives in two worlds. On the one hand, there's the ordinary world, the public world, which he shares with other human beings. On the other, there's a private world where research takes place—a world of passion, exaltation, and despair; a world that is both heaven and hell. The two worlds are more or less closely linked, depending on the individual, depending on the discipline. Jealousy, competitiveness, the need to be recognized are forces that are part of the ordinary world but help to push people into the private world. Likewise, the dreams and triumphs of the private world blend with less glorified discoveries and rewards of the ordinary world.

Physics constructs its representations based on theories, hypotheses, and calculations. These representations are in turn compared with "reality." Quantum theory and the theory of relativity lead to conclusions that contradict our intuitive sense of time and space. Many of the results obtained by calculation make almost no sense translated into the language we use every day. For example, the notion that if you travel long enough and fast enough through the galaxies, you'll become young again. Or the idea that an electron can at the same time be a wave and a particle, that it can simultaneously exist here and there. Or again, at the other extreme of scale, that our universe has a diameter of around (!) ten billion light years. That it was the product of a Big Bang during which an infinitely dense energy abruptly expanded. That our entire world came into being in a few hundredths of a second. That our galaxy moves at a speed of around five hundred kilometers per second. All these notions are derived from complex mathematical reasoning. They make sense only to scientists, who of course don't see the point in translating the equations into everyday language. An immense divide separates the physicist who uses calculations to arrive at a reality that defies the imagination from a public that seeks to grasp a reality whose mathematical evidence belies what their sensory intuition tells them.

The gap between the representation of an object that physics constructs and the one that our senses perceive was aptly described by the English astrophysicist Sir Arthur Eddington[3] in his account of his "two tables." The first, the "familiar table," is a piece of wooden furniture, a flat surface mounted on four legs. It is an everyday object, made of a "substance," that Eddington leans on to write. The second is the "scientific table," "mostly emptiness" with "numerous electric charges rushing around with great speed." To write, Eddington leans the emptiness of his scientific elbow on the emptiness of his scientific table. So, in other words, the representation that physics forms of an object is a theoretical construction. It is the result of hypotheses accumulated over the course of centuries. The world of physics is a world of abstractions, of symbols. That doesn't mean that every symbol represents a precise fragment of the everyday world, or even something that we can explain in terms of sensory experience. It has come to the point where physics has become a little like writing, in the way that the marks of the written word symbolize the thing that the word represents.

Biology hasn't yet reached that point. Biology admits of a great many generalizations, but very few theories. The most important among these is the theory of evolution, because it unites, from a variety of domains, a mass of observations that would otherwise remain isolated; because it links within these domains all the disciplines that concern living beings; because it brings order to the extraordinary variety of organisms—in short, because it supplies a causal expla-

3. Arthur Eddington, *The Nature of the Physical World* (Ann Arbor: University of Michigan Press, 1953), p. xi.

nation of the living world and its heterogeneity. It's curious that theories of physics such as relativity or quantum theory, which are not understood by the public, are neither debated nor contested. With the theory of evolution, it's the reverse. Everyone thinks they understand it. This theory *is* controversial, though challenges to it are often based on reasons that are irrelevant. Biology, with its theories, also attempts to construct a representation of its objects; in this case, of living organisms.

For the biologist, there are two types of organisms—two sorts of dogs, for example. The first, the "familiar dog," is the one we pet and which comes when we whistle to go for a walk. The second, the "biological dog," is a sort of abstract creature, constructed according to current theory and modified as current theory changes. At the start of this century, for example, the biological dog was primarily colloidal. It consisted of a "clone," or collection of cells of various sorts—muscle, nerve, gland, and so on—all produced by the division of the same initial cell, that is, the fertilized egg. The cell itself was seen as a sack of enzymes—the catalyzers of chemical reactions—floating in a drop of "colloidal liquid," a gel-like substance favorable to the action of the enzymes. Installed on the chromosomes, genes determined the properties of the organism, without any connection to the cells. Study of the biological dog was thus carried out in many separate fields that very often were unaware of each other's existence. It was not unusual to find on the same floor of a building, sometimes even next door to each other, a biochemist and a geneticist who did not talk to each other.

At mid-century, the biological dog turned into a molecular dog. Biochemists and geneticists worked in close cooperation on the same organisms, the simplest in the living world: bacteria and viruses. In a few years, the landscape had changed completely. Enzymes were proteins. Each protein was made up of a specific sequence of amino acids. Genes were segments of DNA—the double helix. Each gene was made up of a specific sequence of nucleotides. The gene contained the information that determined the structure of a specific protein. In other words, the sequence of nucleotides in DNA defined the sequence of amino acids of the protein, and thus its three-dimensional structure. The machinery for making protein was minutely disassembled, revealing a two-stage process: transcription of the gene to messenger RNA, translation of the messenger RNA into a peptide chain. Gene expression was modulated by regulatory loops in which various molecules acted to promote or repress the activity of the gene in specific ways.

Biology tells us that, of our two dogs, the true one is the molecular dog. The familiar dog is only a pale reflection, the part that our senses can apprehend. With the sensory equipment bequeathed us by evolution, we can perceive the head, paws, and hindquarters of our dog, but not the cellular clusters or molecular bundles that biology assures us our dog is made of. To perceive only some of these latter aspects of the molecular dog, we would need another kind of sensory equipment, another brain. All the same, if we want to understand how our dog works, where he came from, and how to take care of him when he is sick, it is the molecular dog we have to consider.

In the second half of this century, as we noted, biology was radically altered. But the representation of living organisms current today was formulated in two stages. During the 1950s, the analysis of metabolic processes in particular led to a series of amazing successes. It became evident that foods—sugars, for example—are progressively broken down, in reaction after reaction, into ever simpler molecular fragments. Simultaneously, each reaction liberates chemical energy that can be used for other reactions. Once broken down, the fragments of food in turn become the building materials for the components of the cell. Both processes involve a linear chain of reactions, each step of which is catalyzed by a specific enzyme, a bit like an assembly-line production in an automobile factory, for example. Hence the idea that, in our molecular dog, the biological phenomena correspond to a process of linear transformations in a long series of successive stages. Each enzyme—that is, each protein—presents a structure with a unique amino acid sequence. A protein owes its chemical properties to a specific recognition site that allows it to bind to a particular compound and catalyze its chemical transformation. Now, the amino acid sequence of a protein is coded by the sequence of nucleotides in the gene that determined that protein. So if each protein is unique, each gene is also unique.

Evolution thus took shape as a more or less linear and continuous process that resulted from mutations that altered the properties of proteins. The best adapted among the mutated forms were favored by natural selection. As for the increase in complexity that accompanied evolution, it was the product of the linear and continuous addition of DNA to genomes. This resulted in layers of biochemical and physiological novelties deposited one on top of the other.

That the complexity of the living world could break down into linear molecules and chains of reactions was a satisfying notion for a human brain used to moving in a continuous and linear time from birth to death. But in the 1970s, continuity began to fissure and linearity to fracture. This happened once genetic engineering afforded access to what had remained outside the reach of experimentation: the genes of higher organisms. Now came a harvest of surprises: a break in the continuity of genes; the existence of families of genes—up to twenty or thirty—of very similar structure in the same organism; the remarkable conservation throughout evolution of the structure and function of many genes, which remain almost identical in most organisms. The internal architecture of many genes turns out to be the result of combinations of relatively short fragments of DNA, each fragment coding for a protein motif whose three-dimensional structure and electrostatic charge determine its ability to recognize and interact with other molecules. All of these notions are hardly compatible with the former understanding of the structure and functioning of the molecular dog.

Seen from the new perspective, the structure of the living world is no longer linear and continuous, but, on the contrary, nonlinear and discontinuous. Genes and proteins are no longer unique objects, idiosyncrasies particular to a species. Structures look very much the same from one species to another. Further, within the same species, closely related structures often control very different functions. Moreover, segments with shared sequences are often interspersed with segments of differing sequences. Genes and proteins are, for the most part, sorts of mosaics formed by the combination of a few elements, a few motifs, each bearing a recognition site. There is a limited number of these motifs, one or two thousand. The combinatorial nature of these motifs is what gives proteins their infinite variety. The combination of some particular motifs is what gives a protein its specific properties.

The basic element—the one that's directly involved in the chemistry of the cell—is the recognition site of a protein domain. At first, molecular recognition appeared to be limited to interactions between enzymes and substrates or between antigens and antibodies. Now it is credited with a primary role in a variety of reactions: protein polymerization to form structures such as the proteins found in muscle, the cytoskeleton, ribosomes, viral capsids, DNA-protein interactions that regulate the activity of genes, receptor-ligand interactions in a whole series of phenomena such as signal transduction, cell-cell interactions, cell adhesion, and so on. A number of molecular recognition sites have persisted unchanged throughout evolution in such a way that we find virtually identical versions of them in the most diverse organisms.

These changes also altered the way scientists perceived biochemical evolution. So long as each gene, and thus each protein, was regarded as a unique object, the result of a unique sequence of nucleotides or amino acids, each one could only be the result of a new creation—a highly improbable proposition. But the existence of major families of proteins with identical structures; the formation of "mosaic" proteins out of motifs found in numerous other proteins; the surprising fact that, over the course of evolution, proteins retained their specific motifs and active sites despite enormous morphological diversification—all this shows that evolution proceeded in a very different way from what had been believed. In fact, biochemical evolution seems to function in accordance with two principles: one, the creation of new molecules, the other, their selection.

The creative aspect of biochemical evolution doesn't emerge out of nothing. It consists in making something new out of something old: what I have elsewhere called "molecular tinkering."[4] The very first genes must have been formed from short sequences of thirty or forty nucleotides. These segments then grew, either by attaching to each other or by doubling once or several times. In fact, in many genes we find traces of one, two, three, or even more successive duplications followed by more or less significant diversification. Duplication either of segments of DNA or of entire genes seems to be one of the major modes of molecular tinkering. It is through successive duplication that numerous families of genes formed, such as the hemoglobin genes, and many regulatory factors or genes of the immunoglobulin family that fulfill functions related to regulation, such as antigen recognition, cell adhesion, and axonal guidance.

Another means of gene fabrication is the rearrangement of existing fragments to make mosaic genes. This process also entails an element of selection. The big surprise was to discover the persistence—almost the immutability—of specific recognition motifs in proteins throughout evolution, despite the enormous diversity of species. This persistence was explained by the strong constraints on recognition sites,

4. François Jacob, *Le Jeu des possibles* (Paris: Fayard, 1982).

the basis of all molecular interactions and thus of all the chemical activity of the cell. The need to conserve the specificity of molecular
interactions explains the inertia, across evolution, of the structures in
question. This inertia applies as well to the segments of a gene (i.e.
coding segments, called "exons") that determine the recognition site.
It does not apply to noncoding segments of the gene, or "introns."
Nor does it apply to whichever segment is next to an exon, that is, to
the nature of the segment adjoining the exon in question. Introns and
adjacent segments of DNA can thus be interchanged freely. The result
is a second type of molecular tinkering: matching fragments of
DNA—of exons—to make mosaic proteins.

So, once again, it is a combinatorial system consisting of a limited
number of elements that produced the enormous diversity of structures that form the principal components of the cell. Biochemical evolution depends only secondarily on mutations, contrary to what was
long believed. It is the result above all of duplication of DNA segments
and their rearrangement. Fixed points exist in this evolution—islands
composed of specific recognition sites. Around the DNA segments
that code for these recognition sites, other fragments of DNA trade
places more or less freely, in a kind of ballet. Under these conditions,
the fundamental structures or recognition sites recur in all organisms,
but in contexts that are often different. The whole of the living world
looks like some kind of giant Erector set. Pieces can be taken apart and
put together again in different ways, to produce different forms. But
fundamentally the same pieces are always retained.

The mosaic structure of genes and proteins facilitates multiple interactions. The formation of protein complexes, which can be huge,
adds to these possibilities; specific complexes are employed to carry
out certain fundamental cellular processes, composed of multiple reactions and interactions. This is especially true for operations involved
in cell division or cell-cell interactions or certain stages of morphogenesis. The genes of a set directing these operations are linked by

cellular recognition processes that keep their products intimately associated with one another. The set of genes that governs cell division is the same in yeast as in humans. These genes have kept their function and a good part of their structure throughout the course of an evolution that goes back more than five hundred million years. Antonio García-Bellido calls such sets of genes "syntagms." They function as modules that are used in the architecture of all cells.

Modular construction directed by sets of genes also comes into play in the embryonic development of many species. Perhaps all of them. Organisms, especially insects, seem to develop in the form of repeated segments, that is, in multicellular modules. At first identical, these modules subsequently differentiate in a specific way under the influence of groups of regulatory genes, such as Hom genes. The role of these genes is to modify the rules that direct the development of a standard module. They serve to define a well-delimited region and give each segment a particular identity. Each of these regions or segments is defined by the combination of several Hom genes that function in parallel in the same cells. Similarly, terminal differentiation, which produces the different types of cells observed in the body, makes use of groups of conserved genes that operate in concert, for example, to produce muscle or nerve cells in all organisms biologists have studied, from the nematode to humans.

We see how great a difference fifteen or twenty years can make in our representation of the living world and its inhabitants. It is a profound transformation of our molecular dog. What probably contributed the most to upending some of the old ideas, what perhaps evoked the most surprise in the world of biologists, was the persistence of molecular structures and functions from Cambrian times—that is, across more than five hundred million years of evolution—in the face of the diversity of forms and behaviors of the animal world. Some genes and their proteins have remained more or less intact, with only the minute changes that are inevitable over time. Others multiplied, with minor variations that permitted them to take on new functions. Still others became fragmented, and fragments of varying origin came together to create new structures. Moreover, modules formed by sets of proteins coded by blocks of genes underlie fundamental processes throughout the living world.

It is no longer possible to attribute—as was done for a long time—differences in form and behavior among species to differences in protein structure. Under the so-called new synthesis, in vogue since the 1930s and 1940s, the origin of biological diversity was explained in terms of genetic mutations. These mutations produced variations in enzymes that modified certain stages of embryonic development and, consequently, the form and behavior of species. In a polymorphic population, natural selection favored those protein structures, and thus those genetic structures or "alleles" that permitted the most abundant reproduction. It was by reference to these differences among alleles that the peculiarities found in each species were explained.

For the last twenty years or so, the omnipotence of natural selection and the continuity of evolution have been called into question by the theory of "punctuated equilibria" proposed by Eldredge and Gould.[5] The results of molecular analysis appear to indicate that the formation of new species does not correspond to the simple addition of new genes and new functions to the end development of older organisms. The living world is, rather, like a sort of Erector set. It is the product of a vast combinatorial system in which more or less fixed

5. Niles Eldredge and Stephen Jay Gould, "Punctuated Equilibria: An Alternative to Phyletic Gradualism," in *Models in Paleobiology*, ed. Thomas J. M. Schopf (San Francisco: Freeman, Cooper, 1972), pp. 82–115.

elements—segments of genes or blocks of genes that determine the modules of complex operations—are arranged in various ways. The complexity brought about by evolution stems from new permutations of these elements. In other words, the new forms—new phenotypes—often come out of new combinations of the same elements.

5   Self and Other

....

A scorpion is nervously pacing the bank of a river, looking for a way to get to the other side. A frog comes by. "Would you take me on your back and carry me across the river?" asks the scorpion. "Why? So you could sting me? I'm not crazy," answers the frog. "Why on earth would I want to sting you?" says the scorpion. "We'd both sink. Look, I'll pay you well." Persuaded, the frog takes the scorpion on his back, and begins to swim toward the other bank of the river. But once they are well into the water, the scorpion stings the frog. As it is dying, the frog asks, "Why did you do it?" "Because it's my nature," says the scorpion. And they both sink.

The poor scorpion is a fool. Whatever he thinks, if he does think, and whether he loves or hates his fellow creatures, whether he makes plans or not, his room to maneuver amounts to one alternative: to sting or not to sting. He has no other options. The scorpion has nothing like our "free will." He is a pure product of raw nature.

But let's not be too quick to condemn the poor scorpion. On our scale, and in our own way, we also are constrained by our nature:

> I hear the noise of the city
> And, prisoner without horizon,
> I see only a hostile sky
> And the naked walls of my prison.
>
> Day is done. Here's a lamp
> Burning in the prison.
> We are alone in this cell
> Fair light, sweet reason,[1]

murmured Apollinaire, to remind us that the fair light of the world pursued by sweet reason goes hand in hand with the limitations of our brains. We have no choice but to accept these limitations, but only on condition that our place in the living world (the highest-ranking, of course) be clearly defined. What psychologists call our identity must be as clearly established in the living world as our social or family identity. In a story by George Bernard Shaw, the narrator explains that he had a twin brother. Once, as infants, when they were taking a bath, one of the two drowned. "I never knew," says the narrator, "whether the one who drowned was my brother or me."

This story bothers us, in the same way and just as much as Vercors's novel, *Les Animaux Dénaturés,* bothers us. In that novel, a young ethnologist has come to Africa to study a population which he cannot decide how to classify, whether as apes or human beings. To help clarify the issue, the ethnologist marries a female in the group. A child is born. The ethnologist kills it. He goes on trial in London. Once the court rules whether the death was a homicide or a hunting accident, the matter will be solved. The law will decide nature.

Like every other science, biology has cyclical moods. Spells of optimism alternate with spells of depression. The spells of optimism

---

1. Guillaume Apollinaire, "At La Santé Prison," in *Alcools: Poems, 1898–1913,* trans. William Meredith (Garden City, N.Y.: Doubleday, 1964), p. 203.

correspond with the emergence of a new theory, an original way of looking at living creatures and how they function, which provides new ways of analyzing certain phenomena. The periods of depression result from experimental frustration, when, after a period of exploring the new theory, one finds oneself yet again up against a brick wall. Molecular biology ushered in a period of fantastic optimism, as though all the questions raised since antiquity were suddenly going to be resolved by the magic of the double helix. But the techniques that worked with *E. coli* were ill-suited to the elephant. So a period of depression followed. It seemed that higher organisms would forever remain beyond the reach of the methods that had worked wonders with bacteria. With the arrival of genetic engineering, a new period of optimism opened up. The prospect that we might be able to analyze molecules in detail—and particularly the genes of all organisms—totally transformed our conception of the living world, replacing the one that molecular biology had constructed up until the 1970s.

The history of biology looks a little like a long, chaotic, unpremeditated march toward materialism, reductionism, and unity of composition and function of the living world. For a long time, living beings remained independent, seemingly isolated from one another. Each adult organism was the result of a specific creation, and any resemblances among them were due solely to the whim of the Creator. In the seventeenth and eighteenth centuries, comparative anatomy and physiology discerned, little by little, a few organizational and functional similarities within groups of organisms. By the nineteenth century, however, two sorts of relationships were established among all living beings. First, relationships of construction, since once the existence of the cell was demonstrated, every organism was regarded as a "clone," a colony of cells. Whatever their function—nervous, muscular, glandular, and so on—all cells were made up of similar components, classified under the rubric of carbohydrates, lipids, proteins, or nucleic acids.

In the second category were relationships of descent, with the theory of evolution founded predominantly on relationships between forms, on belonging to certain geologic strata, on comparisons of embryonic development. Until this time, the living world appeared to be a system regulated from outside. Whether fixed since Creation or "perfected" via a series of successive events, organized beings were always composed of a continuous series of forms. The structure of the living world as we know it today expressed a transcendental necessity. That creatures might be different from what they were or that other forms might inhabit the earth was simply unthinkable. The advent of the theory of evolution dispelled any idea of a preestablished harmony imposing a system of relations on organized beings. The necessity for a living world to be the way it was gave way to the contingency that already prevailed in the heavens and for inanimate things. Not only might the living world have been totally different from what it is today, but it might also never have existed. Organisms became the elements of a vast system that included the earth and all that it contained. The form of beings, their properties, and their characteristics were thus subjected to the internal regulation of this system, to the set of interactions that coordinated the activities of the elements.

In the mid-nineteenth century, humans suddenly found themselves an integral part of the living world. Though they were now little cousins of the big apes, they still maintained a genuine sense of superiority over all other organisms. It wasn't just that humans were superior. They were different; other. And even within human society, a certain arrogance among the "high born" and racism among many people have been persistent features. According to the myth of Adam and Eve, all human beings are the descendants of this one couple. That should be enough, in the West at least, to destroy self-importance in genealogical matters, but it prevented neither imperialism nor Auschwitz. Yet a few simple calculations suffice to show that all people alive on earth today are more or less distant cousins, as the English geneticist Ronald Fisher[2] and, more recently, Richard Dawkins[3] have convincingly shown. Often, says Dawkins, we hear people boasting that they are related to the queen of England. But everyone is more or less related to the queen of England. And in actual fact, everyone is related to everyone else. What distinguishes aristocrats from others is only that they are so preoccupied with their genealogy that they take particular care in recording its details.

All of this is not a matter of biology but of basic arithmetic. Biology offers two new ideas that go in opposite directions, thus constituting a sort of paradox. On the one hand, there are several million species, and we think we still know only a fraction of them. Moreover, living species today represent only a tiny fraction of the species that, over the course of evolution, have appeared and disappeared. But this diversity of forms and behaviors hides an astonishing unity of structures and functions. Indeed, it really does seem that all species, from the simplest to the most complex, are still more alike and closer to one another than we believed possible up to now. It is as if evolution always used the same materials but arranged them in different forms. Yet among sexually reproducing organisms, everything is arranged so that all individuals of the same species, with the exception of identical twins, are different from one another. It is as if the entire genetic system at work on the earth had been set up always to produce something different. Hence the paradox: on the one hand, everything that seems so different is, when all is said and done, very much the same; on the other, everything that seems the same is, in reality, quite different.

What underscored the degree, until then unsuspected, of kinship between living creatures was a series of new ideas acquired in the last twenty years. The first concerns the structure of genes and their pro-

2. Ronald Fisher, *The Genetical Theory of Natural Selection* (New York: Dover, 1958).

3. Richard Dawkins, *River out of Eden* (New York: Basic Books, 1995).

tein products. In higher organisms, genes are actually fragmented. The sequences coding the proteins, or exons, are usually interspersed with noncoding sequences, or introns. Proteins are made up of domains. But it's always the same motifs that are used over and over, with minor variations. The genes we find today are thus the result of a combinatorial system that rearranged a limited number of motifs, perhaps a thousand or two thousand. These motifs are themselves derived from a small number of ancient modules.

A second recent contribution: the astonishing developments that have emerged in the last few years from the study of embryos. For a long time, it was possible only to observe the succession of complex events from which the form and structure of the embryo were revealed little by little. Folds were seen to form—"sheets" that slid over one another, coiled up, and unfolded again to generate organs. Within the same family, related species showed remarkable similarities in development. But moving from one phylum to another turned up different types of development. Wasn't that to be expected? Wouldn't generating architectures as diverse as the life forms themselves require very different processes? At the beginning of the nineteenth century, only Etienne Geoffroy Saint-Hilaire managed to recognize a sort of general plan of the body, common to most species, beneath the diversity of forms and development.

At the end of the last century, scientists were no longer content merely to observe the development of an embryo. They wanted to experiment, the way physiologists did. They'd cut a fragment of tissue, remove it, graft it elsewhere, and see what it produced. Each of them employed a favored organism for carrying out these experiments. There were sea urchin fanatics, enthusiasts of frogs (or, better, *Xenopus,* whose eggs are larger), chick devotees, and mouse lovers. Each of the animals had its virtues. Each was particularly suited to certain experiments, but not to others. Each was especially useful for targeting certain aspects of development. But results achieved in one organism more often than not had no significance or connection to what happened in other organisms. The only common elements were certain obligatory stages, like "gastrulation," a sort of early invagination that forms stacked cellular sheets—a critical stage, since that's when different cellular types begin to diverge and certain structures of the new individual begin to take shape. Which is what led a distinguished embryologist to say that the most important event in a person's life is neither the fertilization that begins development, nor birth, nor marriage. It's gastrulation!

But without genetic analysis, there could be no in-depth study of the embryo. And with the sea urchin, the frog, the chick, even the mouse, genetic study remained for a long time either nonexistent or insufficient. At least, that was so until the arrival of *Drosophila* on the embryological scene, which brought about a curious situation. On the one hand, the development of the fly or that of a small, rapidly reproducing nematode could be dissected down to the smallest detail. Hundreds of mutants were isolated and their genes identified. On the other hand, researchers were just marking time. Even work with mice, the model for genetic studies of mammals, and of humans in particular, was limited to producing pure strains and studying long and complicated crosses. Then came genetic engineering. In a few months, the situation changed completely. In *Drosophila* and nematodes it became possible not only to identify genes but also to isolate them, purify them, study their detailed structure, and reinsert them in the organism to see precisely how they function. In organisms for which no extensive genetic studies had been performed, it became possible to get access to some genes. Finally, in the mouse and in humans, about which much was known genetically, it became possible to isolate any gene, to reproduce millions of copies of it in a bacterium, to study it in detail, and to reinsert it in a mouse to identify the stages of development and the tissues in which it was expressed.

Genetic engineering thus brought about a total change in the biological landscape as well as in the means of investigating it. Where it had been possible only to observe the surface of phenomena, it now became feasible to intervene at the heart of things. For the first time, it was possible to access the system that underlay the embryonic development of diverse organisms, in particular mammals. To go on to analyze them as whole entities and in detail is a question of time and work. But now we know that sooner or later we will get there.

The first surprise resulted from comparing developmental genes in a range of organisms. Or rather, from attempting to find out whether genes similar to those known to act as master switches in the fly exist elsewhere. Because the two complementary chains of DNA recognize each other and bind specifically, it is relatively easy to search the total DNA of an organism for the presence of a gene similar to a known gene. For example, there seemed to be little chance of finding the famous Hom genes (the genes that in *Drosophila* establish the anteroposterior axis of the body) in organisms other than insects, because their embryonic development is so different. But geneticists looked around anyway, just to make sure. And wonder of wonders! They found them everywhere. First in the frog. Then in the mouse. Then in humans, the leech, the nematode, in amphioxus and hydra. In short, in every animal examined, a group of genes was found that presented a structure very similar to that of fly Hom genes. Everywhere, these genes seemed to play the same role: that of defining the relative position of the cells along the anteroposterior axis of the animal. If in a mutant fly a homologous mouse gene was substituted for one of these Hom genes, it functioned perfectly and fulfilled the role of the normal gene in the fly. The same is true for human genes.

It is hard to imagine the excitement that rippled through the biological community when these results were announced. It had been known for a long time that many genes and proteins had persisted essentially unchanged throughout evolution, and that certain structures kept their sameness from bacteria to humans. But mostly this was the case with structural proteins, such as those of muscle tissue, or enzymes, such as those that play a role in respiration. That the genes that build the human body could be the same as those that direct the making of the body of a fly—now there was something really unthinkable! Unthinkable, that is, that the same genetic framework might underlie processes as dissimilar as those involved in the development of these two organisms!

Moreover, it is not only isolated genes that resemble one another. A whole system of coordinated elements have been maintained together throughout evolution on the same segment of a chromosome and activated one after another—in the fruit fly as in the mouse—according to a precise order in time and space along the length of the body of the embryo. In mammals—humans and mice alike—the group of genes homologous to the Hom genes of the fly are called Hox and are repeated several times. These Hox genes position the vertebrae, ribs, muscles, and central and sympathetic nervous systems. As in the fly, a mutation in one of these genes causes morphological changes in the future organism and, often, premature death.

Two consequences follow from the fact that these groups of genes recur, more or less complete, more or less repeated, in all organisms studied regardless of their shape and size, and in all phyla. On the one hand, the same kinds of genes determine, in very different animals, the formation of very different structures. Thus we have to conclude that this system serves to establish not specific structures, but relative positions—the axial coordinates—for the cells within the organism. In other words, the function of these gene groups is informative, not structural. On the other hand, the fact that the same groups establish axial coordinates in all animals examined, whatever their developmental processes, underscores the antiquity of the system. In all probability, this system already existed in a primitive form about six hundred million years ago, in some common ancestor of all the animals living on earth today.

It's not only activator genes, those that determine the building materials and the cell chemistry, that are conserved from species to species throughout evolution. It's also the selector or master genes, those that switch on and modulate the activity of the activator genes. Examples of these interactions multiply daily. One of the most spectacular concerns the formation of the eye. The complexity and precision of the structures that compose a human eye—the astonishing qualities of the apparatus that, more than any other, gives us access to the world around us—have made the eye the most common illustration of the impossibility of evolution. This argument is very often brandished against the idea of evolution due to chance. If you are out for a walk and you find a watch, you do not for an instant doubt that it was crafted by a watchmaker; likewise, when you consider an organism of some complexity, with all of its organs, you do not doubt that it was designed by a Creator. How could anyone believe that the mammalian eye, with all the precision of its components, with its geometry, is the result of pure chance?

A diversity of types of eyes abounds in the living world. Obviously, to possess photoreceptors confers a great advantage in many situations. Throughout evolution, the eye developed into diverse forms, several times over, based on distinctly different physical principles. The best known are the lens eye of mammals, that is, ours, and the compound eye of insects, that of the fly. Nothing is less alike than these two kinds of eyes. There is nothing common in their organization, mechanism, mode of development. They were considered to be nonhomologous structures that evolved independently from different prototypic organisms. But that is not at all what has been shown by the recent genetic analyses carried out by Walter Gehring and his group. For several years, mutations have been known to interfere with the development of the eye in humans and in mice. In both cases, the absence of a particular gene causes premature death in an eyeless embryo. In each case the two genes were isolated and analyzed. They are almost identical. Thus, in both humans and mice, these mutations affect the same gene, which has been highly conserved throughout evolution. In both cases, the gene contains two segments with high affinity for DNA, one similar to a segment in Hox genes, the other similar to a segment found in another family of genes called Pax. Once again, it's a question of a master gene, but this time one that controls the development of the eye by acting at some level of the genetic hierarchy. More recently, a gene whose absence prevents the formation of the eye was isolated in *Drosophila*. This gene is almost identical to the one in the mouse. So we must conclude that this regulatory gene is essential to the development of the eye in both insects and mammals. Once again, the result created amazement. It ran counter to everything written in the textbooks. Indeed, it had seemed clear to everyone that the compound eye of insects and the lens eye of mammals are totally unrelated structures that evolved independently. Now it appears that both derived from a common prototype.[4]

No less astonishing is the demonstration that this gene by itself commands the entire hierarchy of regulatory events involved in the development of the eye. As we have seen, genetic engineering actually managed to insert this gene in a fly in such a way that eyes appeared on the wings or on the legs of the insect. And the same amazing result occurred when the gene that was inserted was not that of the fly, but of the mouse.

These studies on the eye thus yield two new and startling observations. First, it appears that the activity of a single master gene suffices to initiate the entire cascade of events and structures that make an eye. It probably takes a few hundred genes to manufacture a compound eye or a lens eye. But it's the activity of that *single* master gene

4. R. Quiring, U. Waldorf, U. Kloter, W. Gehring, "Homology of the *Eyeless* Gene of *Drosophila* to the Small Eye Gene in Mice and Aniridia in Humans," *Science*, 265 (1994), pp. 785–789.

that initiates the entire cascade, the entire hierarchy of structures. This is one of the few cases known, if not unique. Indeed, it is a little surprising, because it carries a risk of error. A whole range of mishaps can activate the master gene and unleash the process that leads to the formation of an eye under erroneous conditions of space and time— the kind that can make eyes sprout on wings or legs. The other point is that we never tire of being surprised by the spectacle of nature, which, over and over, uses the same genetic elements to construct very different organs. And in the decades-long battle between "holistic thinkers" and "reductionists"—those who want to study the whole and those who prefer to study the parts—it does seem that the latter draw ever closer to victory.

All living beings, from the most humble to the most complex, are therefore related. The relationship is closer than we ever thought. With the same elements—the same units—the living world diversified endlessly over the course of evolution. It is as if, always under threat, life had to take the most diverse forms to maintain itself, had to make use of the most varied behavior, to occupy the most remote corners of the earth. And diversity here concerns not only differences among species, but also between individuals of the same species. And this is where the second part of the paradox comes in. For, over the course of the last twenty or thirty years, biology has increasingly elucidated what characterizes each individual in the most diverse sexually reproducing species, ours in particular. Immunological differences were first revealed by skin and organ grafts, then by the study of the genes that determine the structure of the molecules on the surface of cells as well as those that direct mechanisms of rejection. Multiple genetic differences have been highlighted by comparing the DNA of various individuals, which led to a definition of genetic individuality, more distinctive than fingerprints and better for identifying criminal perpetrators or establishing paternity. Immunology and genetics have thus amply demonstrated that, with the exception of identical twins, each of us is different from all the other human beings who ever lived, now live, or will live on earth.

These differences between individuals are of particular significance in pathology and medicine. For a long time physicians believed that many diseases depended on two kinds of factors: either exogenous factors, such as microbes, viruses, food, poisons, toxins, and so on, often well defined and catalogued; or endogenous factors, much less distinct and generally clustered under the rubric of "susceptibility" or "predisposition" and manifested by the fact that not all people are susceptible to the same illness in the same way. With progress in human genetics it became apparent that this idea of susceptibility— the fact that a person has a greater tendency to be affected by some diseases than by others—in the end reflects the genetic constitution of that person. In particular, analysis of the chromosomal region containing the human lymphocyte antigens (HLA), especially the work carried out by Jean Dausset's team, showed that individuals carrying certain combinations of HLA genes were more susceptible than others to a particular disease. For example, ankylosing spondylitis, a painful and debilitating disease of the vertebral column, is almost one hundred times more likely to appear in patients harboring a certain genotype than in people who don't.

For about fifteen years, the joint efforts of genetics and molecular biology have enhanced our ability to analyze the genetic makeup of human beings. For a long time, microbiologists and physicians chased after the bacteria and viruses they believed responsible for infectious diseases. Today, geneticists and physicians hunt for the genes that they suspect play a role in hereditary disorders. The successes are mounting. Not only are new genes identified daily, but they are localized on chromosomes, isolated, their DNA sequenced, and the means are being developed to test for them on anyone. The work has led to the discovery of molecules whose existence until now was totally unknown, like the regulator of transmembrane conductance whose lesions appear to be the cause of cystic fibrosis.

Up to now, the physician called to the bedside of a patient made a diagnosis, and on that basis he tried to predict the evolution of the illness in the form of a prognosis. Now he looks to evaluate the structure of genes, the susceptibilities and predispositions, on the basis of which he predicts the future state of the patient's health. Moreover, predictive medicine does not merely evaluate the future health of our fellow citizens, the men, women, and children who presently live around us, whom we might meet on the street. Medicine is also concerned with the next generation, those who are preparing to follow us tomorrow. In fact, medicine does not limit itself to treating the organism after birth, as it did for a long time. It uses all the means at its disposal to examine an individual's condition as soon as possible after conception. In this way it also tries to predict what the future child and future adult will be like. It attempts to specify the person's condition, organs, shape, and potential defects. For a long time, the means of study were limited, like the mode of examining a patient—palpation, percussion, auscultation. Then came X rays, which made it possible to see more clearly, but which soon after turned out to be hazardous to the future health of the fetus. Recently, physicians have gained access to a whole arsenal of complex instruments, for instance, ultrasound and magnetic resonance imaging, which enable them to see the fetus with a precision and clarity previously unknown.

But what really advanced analysis and prediction were the methods that made it possible to sample fetal tissue. These samples are made by amniocentesis or by biopsy of the trophoblast—the tissue that constitutes the external wall of the egg at the point where it makes contact with the uterine mucous membrane of the mother. These methods, which are not entirely without risk for the fetus, gradually improve. Only a few fetal cells are needed to be able to make a prenatal diagnosis, that is, to analyze the state of the genes considered important in a given case.

Today we know what genetic anomalies are at fault in more than fifteen hundred hereditary disorders. There are DNA probes for analyzing the properties of many of these genes. Other genes can be tracked across generations because they are linked to polymorphic markers.[5] We are beginning to be able to evaluate predisposition to diverse diseases, especially to certain forms of cancer. With the work going on in many laboratories around the world, with the advances in establishing chromosomal maps and DNA sequence maps of the human genome, it is clear that we will gradually identify more and more genes whose lesions are implicated in diverse aspects of pathology. The stage is thus set to forecast the part that biology will play in shaping people's destinies, depending on their genetic constitution and, in particular, the structure of certain genes known to have possible pathological effects. But it is important to note that although a growing number of malformations of genetic origin have been identified, we don't, unfortunately, know how to treat all of them. We still don't understand the precise effects of many genetic lesions, and we currently have no way of correcting these defects. Other pathological conditions, by contrast, can be cured or improved by diverse means. Gene therapy, which is only in its infancy, will most likely bring a solution to certain hereditary diseases and cancers.

The predictions that gene diagnostics leads us to make about the fetus or young person are of quite a different kind. To begin with, the diagnosis may be one of a disease that will certainly occur, the genetic lesion leading inexorably to a well-defined pathology. The disease usually first appears at birth, or shortly thereafter, as in hemophilia, congenital malformations, and certain errors of metabolism. With other

5. Segments of DNA that vary among individuals. [Trans.]

diseases, by contrast, the person lives a perfectly normal life for a long time, the pathological manifestations appearing only later. This is the case, for instance, in Huntington's chorea, a fatal neurodegenerative disease whose effects only show up at about age forty, or polycystic kidney disease, or Alzheimer's. These are time bombs, as it were, buried in the genome. They remain perfectly silent before exploding in the prime of life.

In all of these diseases, the lesion of one or several genes suffices to unleash a pathological state. In other conditions, however, the genetic lesion alone is not sufficient. It predisposes to the disease. It increases the probability of contracting it. But for the disease to occur, other events must contribute, most often of environmental origin. Such is the case in ankylosing spondylitis, which I've already mentioned. And there is an entire series of predispositions tied to one or another haplotype HLA, such as, for example, childhood insulin-dependent diabetes or idiopathic hemochromatosis, a fairly serious disease of iron metabolism. Likewise, people can have predispositions to one cancer or another, and we are beginning to be able to specify the genetic constituents in these cases. Thus, for example, it can be predicted that a given person will have a high probability of contracting colon cancer in his life, but not skin cancer or lung cancer.

So what does this bit of knowledge of the future bring us in the domain of health? Until now, confronted with a patient, medicine established a diagnosis from which it drew a prognosis. Now it evaluates the genetic profile right away, from which it predicts the medical destiny of the person. We no longer interrogate the gods to learn about a person's future life or that of his descendants. We interrogate the genes. As always, with the novelties that research brings, both the best and the worst can emerge. The best, because forewarned is forearmed if a treatment or a way of life will provide escape from an unfortunate genetic legacy. For example, prenatal diagnosis of phenylketonuria, a metabolic disturbance that threatens the developing baby with mental retardation, can be countered with a nutritional diet for the newborn. Or despite a confirmed predisposition, diabetes can be prevented from appearing by means of appropriate measures, or rheumatism from becoming established too soon or too suddenly by avoiding intestinal infections.

But this predictive power can also be for the worst when people have no means of combating the disease that is predicted——when, for example, prenatal diagnosis shows that the child to be born will be afflicted with a fatal disease such as thalassemia, or illnesses that involve crippling disabilities, such as severe myopathies. These are dramatic situations that could compel an expectant mother to consider terminating a pregnancy. Another worst-case scenario is one where the genetic diagnosis shows that a young man or a young woman in full health carries the dominant, inescapable gene for Huntington's chorea, for which no correction or treatment is yet known. Interrogating the genes thus comes back to asking a series of difficult questions. Do you want to know when and how you are going to die? Do you want to know how you are going to react to such news? Whom would you give access to this information? Your family? Your boss? Your insurer? The state?

All of this means that, in order to progressively turn the worst into the best, to find treatments where none yet exist, it is necessary to carry on research unceasingly. The developments that we rightly expect from gene therapy, still in its infancy, can only follow this direction.

Medicine is clearly undergoing an evolution, especially with the development of the analysis of the human genome. Indeed, thanks to this analysis, most, and in the end all, of the pathological conditions created by the lesion of a single gene will eventually be recognized. Prenatal diagnosis and identification of late-onset disorders, such as

Huntington's chorea, before symptoms appear will weigh heavily on afflicted patients who otherwise would never find out or be aware of their condition. Until now, a person became "sick" only after symptoms appeared. People would go to the doctor complaining of a few aches and pains. With the availability of the data on the genome, future illnesses or the risk of illnesses will be revealed. We already know about families in which autosomal dominant disorders such as Alzheimer's, colon cancer, and breast cancer wreak havoc. Without a doubt, a whole series of pathological conditions will cluster under this category. People will become patients before their time. Their condition, their future will be discussed in medical terms, even though they feel fine and will remain in good health for years.

At least for monogenic Mendelian lesions, the data on the genome will convert the potential into the actual. Even if the illness is not yet manifest, its biochemical proof will have been documented. The knowledge of the genome will lead to the identification of a growing number of DNA sequences in which certain changes are associated with an accrued risk of pathological states such as diabetes, depression, cancer, cardiovascular disturbances, and so on. In certain cases, the pathological state will depend on the combination of several genetic modifications. In other cases, environmental factors will come into play. But we will consider people with markers for cardiac disturbances, schizophrenia, or cancer as already sick, even before they have any pathological lesions. The prediction of clinical symptoms in an individual will remain relatively imprecise, inasmuch as it depends on the study of populations. But the reality of risk will become much more tangible. Today, risk is measured by abstract numbers that barely have any effect on a person's perception of himself. In future, we will know how this same risk is recorded in chemical terms, in a person's genome, as an indelible part of himself. Potentially sick people who hold positions of power will be closely monitored. They, as well as their physicians, will watch out for the appearance of symptoms.

Whether means of treatment exist or not, potential disorders will in the future announce their presence as never before.

Thus we see how the dialectic of Self and Other has altered, how it has changed nature. We are at once all close relatives and all different. Sexual reproduction is in actual fact a machine for making Other. Other than parents. Other than all the other individuals of the species. But the diversity produced this way is not always well accepted. It is rarely taken for what it is, a motor of evolution. Too often biological diversity and social or cultural diversity are confused. Some people invoke biological diversity to rationalize inequalities in the social order; they justify inequalities by reference to a supposed natural order that classifies others according to "norms" that they themselves have fixed. Or else biological diversity is considered a scandalous topic by those who criticize the social order and would like to see everybody as identical. Consider, for example, the oft-heard but nonsensical phrase "inequalities regarding the disease." Inequalities are a function of treatment, not a property of the disorder. We can speak of *differences* concerning diseases, not of inequalities. Otherwise we confuse two distinct notions: identity and equality. The first applies to characteristics—physical as well as mental—of people, the second to their social or legal status. The former is dictated by biology and education, the latter by morality and politics. There is no equality in biology. Molecules and cells are neither equal nor unequal. It takes the humor of George Orwell to remind us that animals are more or less equal. But too often this confusion is used in the service of political and social ends, either by imposing equality on identity or, conversely, by trying to preserve inequality by justifying it with differences. But it is precisely because human beings are different that the concept of equality had to be constructed. Because there are strong people and weak people, clever people and idiots. If we were all the same, the idea of equality would be uninteresting.

Diversity is found at the very roots of biology. The genes, which constitute the heritage of the species, join and separate with the passing of generations, forming the ever different, ever transitory combinations that are individuals. It is this infinite combinatorial system of genes that makes each of us unique. It is what gives the species its richness and its variety.

6   Good and Evil

....

According to the myth of Adam and the story of Faust, knowledge can only lead to evil, symbolized by the serpent and by the devil, respectively. In *Antigone* by Sophocles,

> With some sort of cunning, inventive
> beyond all expectation
> [man] reaches sometimes evil,
> and sometimes good.[1]

The twin myths of Prometheus and Pandora are more subtle. If we take Hesiod's word for it, the fate of humanity was played out in the battle that pitted Prometheus against Zeus, knowledge against power, the rational against the irrational. In fact, it was Prometheus who created man with clay and water. Athena had taught him astronomy,

1. Sophocles. *Oedipus the King; Oedipus at Colonus; Antigone.* 2nd ed., trans. David Grene (Chicago: University of Chicago Press, 1991), p. 175.

mathematics, architecture, navigation, medicine, and "other very useful arts." He used all his knowledge in the service of humankind. He was, says Albert Camus, "the hero who loved men enough to give them fire and liberty, technology and art."[2] But he very soon clashed with Zeus, who detested men as too ambitious and too pretentious for his liking.

Prometheus was punished and chained to his rock because he transgressed the law of Zeus and tricked him for the sake of mankind. Drawing all the rest of us into his misfortune, Prometheus was caught in a trap that he himself had set, while humankind was punished through the agency of Pandora, the woman created by Zeus to deliver a box containing all the world's evils. If she had not opened the lid, allowing these evils to spread throughout the world, men would have continued to live as before, "protected from suffering, hard labor, and painful sicknesses that bring death." A response to the trickery of Prometheus, Pandora was herself a trick. She was deceit made woman.

The themes of Prometheus and Pandora represent two aspects of the same story: the origin of human misery. Man's need to toil on earth to earn his bread comes to the same thing as his need to reproduce himself through woman, to be born, to die, to feel each day the fear and hope of the next. According to Jean-Pierre Vernant, Pandora introduced a fundamental ambiguity into the world. She brought mixture and contrast into human life. From now on every good would be twinned with its evil counterpart, every light with its shadow. Zeus willed that good and evil, born of knowledge, should not only be mixed but inextricably joined together—inseparable.

That's just what we often observe today. Some of the ill effects

2. Albert Camus, "Prometheus in the Underworld," in *Lyrical and Critical Essays,* ed. Philip Thody; trans. Ellen Conroy Kennedy (New York: Knopf, 1968), pp. 138–139.

brought about by science and its applications resulted from good in-
tentions. The first radiologists, for example, hardly suspected that X
rays could cause cancer. Nor chemists that fertilizer aimed at improving crop yields would be the cause of dangerous pollution. Nor
physicians that the generalized use of antibiotics would result in the
selection of drug-resistant microorganisms. And no one could have
suspected that the speed and scale of growth that medicine and public
health have experienced since the end of the nineteenth century
would lead to overpopulation, which poses one of the gravest threats
to our planet.

Prometheus symbolizes mankind's battle against nature, against the
natural order established by the gods. Humans have never ceased to
fight: against poverty; against cold; against disease and death; against
the violence of the world around them. Trapped in his mortal condition, man refused to bend to the laws of nature. He refused to be
an animal or to be only an animal. He expressed this refusal from the
very beginning, from the invention of fire, writing, and arithmetic.
Science came rather late to this struggle, providing ammunition. As
a matter of fact, the history of science is in a way the history of the
battle of reason against revealed truth.

Modern science originated and developed according to a conception the Western world inherited from Greek culture: that of speculative knowledge based on a criterion of truth. This criterion of truth
depends on the correspondence between the representation, as expressed in discourse, and reality. This speculative knowledge provides
an adequate view of the world, and accurate description is itself the
ultimate aim of knowledge. Learning about the world permits us to
grasp the most profound aspects of reality: its principles and its origin.
Hence the idea that scientific theories replace each other in succession, moving closer and closer to an ideal theory, one that will give a
definitive representation of reality. "Science is the asymptote of

truth," said Victor Hugo. "It approaches unceasingly and never touches."[3]

But in the course of this century, the nature of science—at least experimental science—changed. It was no longer a simple way of knowing, a body of knowledge. It became a major sociocultural force directing the destiny of our societies. If science exercises such a profound influence nowadays on social life, which it has modified even down to our systems of values, it is not only because of the new view of reality it offers us. It is also, and especially, because it has produced a set of scientific practices, techniques, and machines that have transformed our way of life. As a matter of fact, the classical demarcation between science and technology has gradually become blurred. There is very little difference these days between a university research laboratory devoted to so-called basic research, and an industrial laboratory interested in possible applications of its findings. Both cases illustrate directed research that pursues well-defined objectives and calls upon a socially well organized activity. The question is no longer simply one of decoding the world, but also of transforming it.

Moreover, owing to progress in physics and biology, research demands more and more elaborate instrumentation. Consequently, it requires increasingly powerful high-technology industrial support to produce the needed apparatus. Apparatus which, when all is said and done, is only the translation and practical expression of a body of scientific theories. In the resulting play of interactions—a complicity between science and technology—the advances of one depend on the advances of the other, and vice versa. However, science and technology are not identical; their interests, their rules of functioning are different. One aims to produce knowledge, the other to act upon the world. The first

3. Victor Hugo, *William Shakespeare,* trans. A. Baillot (Boston: Estes and Lauriat, n.d.), p. 96.

endeavors to interpret, to understand; the second to dominate, to master. Though it is often necessary to dissociate them, they complement and nourish each other. This new aspect of science in close liaison with an ever expanding and ever more rigid and determinative technology is what so profoundly affects today's social life and culture.

To become convinced of this influence, we only need to consider some of the effects modern biology has had on Western culture. On ideas, first of all, then on some questions raised by the new kinds of interventions made possible by the new technologies. The achievements of modern biology have led to certain notions that often run counter to long-held ideas, some of which are still current. The living world is characterized both by obvious diversity and by hidden unity. There are sperm whales and microbes, fleas and giraffes, organisms that thrive at high temperatures and others that thrive in frozen regions. But if we look beneath the variety of forms, we find an amazing similarity of structure and function. Can we imagine a better demonstration of the theory of evolution? Many of the same compounds or chemical reactions turn up from microbes to mammals. The difference between a fly and an elephant, an eagle and an earthworm, is due not to changes in their chemical constituents but to the distribution of these constituents. In all vertebrates, we find the same chemical reactions. It's not the differences between molecules that make one mammal different from another. It's the modifications, often minor, that arise in the course of embryonic development.

Another idea highlighted by biology is the importance of diversity in the living world. Diversity of species on earth, of individuals within species. It's diversification of individuals, their gradual or relatively sudden divergence, that underlies the formation of new species. By diversifying in the extreme, by forming millions of new species, living forms little by little occupied our entire planet, invading every nook and cranny, all possible niches. A subtle play of genetic mechanisms, the main one of which is sexuality, contributed to this diversification.

Sexuality is a veritable machine for making things different; it makes each organism, with the exception of identical twins, unique. It makes each individual, animal or human, different from all others now living, those that have lived, and even, probably, those that will live. Genetic diversity, which is responsible for the richness of animal and plant species, contributes in the same way to the human species. It is both the result of and the motor that drives biological evolution. For the human species as a whole, and for each population, it constitutes a great asset. The immense variety of physical and mental aptitudes gives human populations their plasticity, their ability to respond to new environmental challenges, their potential for adaptation and creation. A population composed of genetically similar individuals would be at the mercy of accident—epidemics, say, or abrupt changes in circumstances. Any effort to homogenize the biological properties of individuals, either out of desire to "better" them through eugenics or to enhance certain aptitudes—say for mathematics or running— would be biologically suicidal and socially absurd. For the group as well as for the species, the genetic value of the individual does not reside in the specific quality of his genes. It is that he doesn't have the same collection of genes as others do. It is that he is unique. The success of the human species is due, among other things, to its biological diversity. Hence diversity among human beings has to be carefully preserved. Particularly since cultural diversity, which played an even more important role than did genetic diversity in the development of humankind, is today seriously menaced by the model industrial society has planned for the future.

We owe all of this new information in large part to molecular genetics. Since the discovery of DNA by Watson and Crick, molecular genetics has come to occupy a central place in our system of interpreting and explaining the living world. Genetics analyzes the blueprint of the future organism—a blueprint contained in a series of genes transmitted by the germ line cells—and defines its architecture.

For a long time, our understanding of an organism's internal structure was founded on the methods of classical genetics. Following the behavior of characteristics across generations, these methods allowed us to identify and localize the genes on a chromosome map. The experiments involved research using mutants, having them reproduce, and varying the combinations of crosses. These procedures could only be applied effectively to microorganisms or to little multicellular organisms whose life cycle is very short. In the last twenty years, new techniques of molecular genetics have brought entirely new analytical methods. The possibility of cloning DNA, amplifying it, sequencing it has obviated the constraints imposed by the cycle of reproduction of organisms or the limits imposed by breeding. It has become relatively simple to do a genetic analysis of any organism, in particular of humans and others that the methods of classical genetic analysis excluded.

Thanks to these new methods, it has become possible to trace how genes are conserved or modified in the course of evolution. And also to observe how evolution goes about creating new molecular structures. We can even pick out the main tricks evolution uses to produce the new from the old, as if it were a tinkerer who, over millions and millions of years, slowly rejigs his work, repeatedly touching it up—cutting here, lengthening there, seizing every opportunity to adjust, to transform, to create.

These new ideas concern the status of humankind or, rather, of individual persons, all related, all different. They also concern the relations humans have with other species, as well as the tinkering nature of evolution. Of course, this entire transformation of our ideas about heredity and kinship could not help but clash with the traditional views held in Western culture. Perhaps an even greater clash is that caused by modern biology's view of reproduction and sexuality—the possibility of dissociating these two processes, of controlling birth, of "artificially" fertilizing women with frozen sperm, of obtaining fertilization in vitro, of permitting an embryo to develop in a uterus other than its mother's, and so on. For reproduction and sexuality are not only installed in the core of living beings. In our species, they are also situated precisely at the point where culture and nature meet.

Although ethnologists debate themselves breathless concerning the origins of the incest taboo, they do agree on its universal character. "After all," writes Alfred Kroeber, "if ten modern anthropologists were asked to designate one universal human institution, nine would be likely to name the incest prohibition; some have expressly named it as the only universal one."[4] To explain the incest taboo, some people have invoked exclusively natural causes. Others see in it a phenomenon whose origin is purely cultural. Today, most anthropologists agree that the taboo has roots in nature as well as culture.

For Claude Lévi-Strauss, this is the intersection of nature and culture. "In one sense, [this prohibition] belongs to nature, for it is a general condition of culture. Consequently, we should not be surprised that its formal characteristic, universality, has been taken from nature. However, in another sense, it is already culture, exercising and imposing its rule on phenomena which initially are not subject to it."[5] For the anthropologist, the incest taboo and exogamy play an essential role: among humans, they define relationships without which they could not raise themselves above biological organization to achieve social organization.

The idea of life has played an important role in the great myths and great religions. Almost all traditional cultures have felt the need to exalt living beings. In those cultures, life is always imbued with magic

---

4. Alfred Louis Kroeber, "Totem and Taboo in Retrospect," in *The Nature of Culture* (Chicago: University of Chicago Press, 1968), p. 307.

5. Claude Lévi-Strauss, *The Elementary Structures of Kinship,* trans. James Harle Bell, John Richard von Sturmer, and Rodney Needham (ed.) (Boston: Beacon Press, 1969), p. 24.

to some extent. A sort of fetishism attaches to it. Living matter possesses miraculous properties—it is activated, influenced, transformed. With their procession of images, metaphors, affinities, living beings occupy a privileged place in the world. They automatically rank above all other entities; they are always accorded the most weight. Compared with living beings, inanimate objects appear colorless and flat. From things to beings, from dust to thought—this is a hierarchy of value as much as of complexity. Phenomena are not only more complicated in living beings, they are also more perfect. Quality without peer calls for causality without peer. Perfection becomes an explanatory principle. The need of traditional cultures to exalt life in general, and human beings in particular, also expresses the exceptional relationship that these cultures assume binds living beings directly to the forces that govern the world. Indeed, life is considered to be sacred. It testifies to divine intervention. Only divinity can give life and, consequently, take it back. Parents have no other function than to carry out the divine will.

So long as we remained in almost complete ignorance of the mechanisms that underlie life, we had no choice but to attribute them to supernatural principles. But from the moment we began to know something about the reactions that caused them, from the moment that we were able to discern even the possibility of intervening at the molecular level, the old values became controversial. New knowledge led us to question not only traditional representations but also norms based on the very high value and sacred character assigned to natural forces. It became more and more difficult to maintain a systematic and unquestioning respect for processes rooted in nature.

The techniques of molecular biology provide access to the material underpinnings of heredity. They allow us to tinker with DNA, to cut it at precise points, to join sequences together—in brief, to achieve in the laboratory the kinds of manipulations that evolutionary tinkering effects in nature. What the species barrier prohibits, biologists have found a way to accomplish at a more profound level. These kinds of manipulations, designated by the term "genetic engineering," gave biology a tool of a previously unimagined power. Genetic engineering brought an entirely new experimental access to the study of questions as complex as cancer, or brain functioning, or the development of the embryo. Today it has become an indispensable instrument in most areas of experimental biology.

But genetic engineering has also incited passion and hostility. It has even become one of the principal sources of mistrust of biology. Not so much because of its dangers, which have been debated and are no worse than those scientists overcame long ago in experiments on bacteria and pathogenic viruses, but simply because the idea that we can take genes from one organism and insert them into another upsets us. The notion of what we call "genetic manipulation" or "recombinant DNA" seems to us to border on the supernatural. It suddenly recalls, from the mists of time, some of the myths rooted in people's fears. It arouses the terror that visions of monsters provoke in us, the loathing associated with the idea of hybrids, beings merged in defiance of nature.

For convincing evidence of such fears we need only to look at the depictions of the Last Judgment rendered over the course of the centuries. The ones by Hieronymus Bosch, for example. The Hell he shows us is peopled with the most frightening monsters he could imagine. And the most horrible monsters, those charged with torturing sinners, are precisely unnatural hybrids: revolting mixtures of fish and dog, rat and insect, man and bird. For Bosch, the best way of instilling fear was to juxtapose the disorder of an imaginary world onto the order of our everyday world. It is the old nightmares that genetic engineering experiments reawaken. Genetic engineering evokes maleficent knowledge; forbidden knowledge; the very model of knowledge we must not acquire. Bosch reminds us of Prometheus, punished for having stolen the fire reserved for the gods. Most out-rageous is that it is so easy to tinker with the substance that is at the very root of life; that it is so simple to play with what is still the most marvelous story and the most disconcerting problem in the world, which is the formation of a human being: the process that, following the union of a sperm and an ovum, unleashes the division of the egg, the production of two, then four cells; then a little ball of cells, then a little sack. Meanwhile, cells appear in this potential little being that gradually form a little mass of nerve cells. And these cells will allow the being to speak, to write, to count, to play the violin, to cross a street between cars, to paint, to write a book. Inside this little mass of cells are algebra and music, syntax and semantics, geometry and counterpoint. Is it possible to imagine a more fantastic story?

Genetic engineering has come under the most serious criticism that can be made against science: that it gives biologists the power to debase and subjugate both the human mind and body. As a matter of fact, the power to alter human beings is not new. Several thousand years ago it was already possible to set up a program of selection among humans and to create strains as different from one another as Pekinese and German shepherds or Great Danes and basset hounds. Breeding and selection were, in fact, invented by prehistoric farmers. They are empirical arts, applicable to humans as well as to horses and cows. The application may even work better in humans, whose bodies are not specialized like those of horses, birds, or fish, and who, in the animal kingdom, rank as dabblers. Human beings present a remark-able diversity of properties and thus an immense evolutionary poten-tial when subjected to different selective regimes.

Selection among humans is precisely what was advocated by a cousin of Darwin's, the Englishman Francis Galton, who in 1883 cre-ated the word "eugenics"—literally, "good genes." "It would be quite practicable," he wrote, "to produce a highly-gifted race of men by judicious marriages during several consecutive generations."[6] Galton was a curious character. A biologist and statistician, he possessed a very brilliant mind. To clarify the role and effectiveness of prayer, he compared the average age of death for those for whom people pray the most—the kings of England—with that of ordinary people. Failing to find any difference, he concluded that prayer was useless. Galton passed a good part of his life analyzing and comparing the hereditability of certain traits, both physical and mental, among a variety of individuals and populations. He deduced from his data that all the traits he studied were hereditary: intellectual capacities as well as physical properties, talent as well as mental indigence, madness and even poverty! For him, the processes of evolution and natural selection functioned seamlessly, some pushing toward the deterioration of the human species, others toward its improvement. "Our part is to watch out for opportunities to intervene by checking the former and giving free play to the latter" (p. xxvii). Hence the definition of eugenics as the science that allows the human species to improve by giving the best "races"—the best blood—a greater chance to overtake inferior races.

The idea of improving human reproduction is an old story. It goes back to antiquity. Plato wanted to limit births among the poor, whom he thought unintelligent. At Sparta, newborns whom a committee of elders judged to be malformed were thrown from a high cliff. There are still cultures today that do not value girls and which, at birth, get rid of female babies, or at least some of them. This negative eugenics—by elimination—is the obverse of positive eugenics—by selection—in the attempt to generate the most satisfactory babies possible. Actually, reproduction is never completely unchecked. Each culture

---

6. Francis Galton, *Hereditary Genius: An Inquiry into Its Laws and Consequences* (London: J. Friedmann, 1978 [1869]), p. 1.

has rules that proscribe certain types of unions, often tied to incest. In our Western societies, marriages between very close relatives are forbidden. As new hereditary diseases are recognized, "genetic counseling" that aims to dissuade certain couples from reproducing takes on greater significance. Abortion becomes more acceptable. In Europe, both methods were used in the effort to eradicate thalassemia. In Greek villages, where there is a high rate of heterozygotes and where marriages are still often arranged, attempts were made to make prenuptial diagnoses, which were kept generally secret, in order to prevent marriages between heterozygotes. The only result was to prevent all marriages between heterozygotes, who were quickly spotted by the population. In Sardinia and Cyprus, eradication efforts began later, when it became possible not only to identify heterozygote carriers but also to make prenatal diagnoses of thalassemia. In these cultures, people preferred to marry freely and to abort in the case of a homozygote fetus. The result was a considerable reduction of the disease.

In these countries, negative eugenics appeared more tolerable, more manageable, seemed to encounter fewer obstacles than positive eugenics. And yet many well-intentioned people think that within a society that promotes the rights of the individual, it is in principle possible to improve our posterity with the knowledge and know-how already available, with measures such as a system of controlled or "recommended" unions and adjustment of the number of children authorized for each couple. The use of frozen sperm from carefully chosen donors has been advocated. Some people are even excited by the idea of sperm from Nobel laureates (since they probably don't know any Nobel laureates!). But how should we select for complex traits directed by polygenic systems about which we know nothing? With dogs or with cows, we know the characteristics we want to select. But with humans? Discounting factors that govern simple traits and follow Mendelian segregation, out of context, which genes are we to consider the best? And then, said Bernard Shaw, "what has posterity done for me that I should do something for it?"

At the beginning of the century, Galton's eugenics was well received. Most geneticists looked favorably on the theory, among them some of the biggest names: Morgan, Fischer, Haldane, Muller, and others. Several even sketched out scenarios that aimed to improve the common gene pool. Eugenics societies were created in England and the United States. In the United States, programs were begun to sterilize several thousand people described as "mental defectives" over a period of about twenty years. All these scientists who had promoted eugenics, who had developed it into a theory and proposed ways to use it were no doubt sincere. They believed in the soundness of their science. They wanted to use it in the service of humankind. They didn't count on Hitler.

It is hard to believe that the racist ideology of the Nazis wasn't fed by ideas about eugenics that dated from the beginning of the century, as Daniel Kevles's book argues convincingly.[7] Among the theory's influential proponents was the American geneticist Charles B. Davenport, who founded Cold Spring Harbor Laboratory to study human evolution. Irresistibly drawn by eugenics, Davenport wanted to protect the white population in the United States from what he considered genetic pollution by blacks, Poles, and Italians. He was president of the International Federation of Eugenic Organizations. In that capacity, he asked his friend Eugen Fischer, professor of anthropology at the University of Berlin and the best human geneticist in Germany, to preside over the Research Committee on Racial Cross-breeding. Fischer was also one of the authors of the manual on *Human Heredity and Racial Hygiene* from which Hitler, in prison, had nourished his racism. Elected rector of the University of Berlin, Fischer expressed

7. Daniel J. Kevles, *In the Name of Eugenics: Genetics and the Uses of Human Heredity* (Berkeley: University of California Press, 1986).

satisfaction at the intervention of political power in the life of the country exemplified in a biological population policy aimed at eliminating inferior beings. During the same period, Konrad Lorenz compared the elimination of individuals who are asocial by reason of their deficient constitutions with the elimination of a malignant tumor, an operation that seemed to him easier and less risky in the first case than in the second. From Fischer we get to his student and successor, the professor Count Otmar von Verschuer, specialist in internal medicine. And from him to his assistant, the notorious Dr. Joseph Mengele, SS captain and physician of Auschwitz under the authority of Verschuer.

The German geneticist Benno Müller-Hill[8] has described how Verschuer and Mengele worked with the best scientists of Germany. Their so-called research was conducted within the official scientific institutions. It benefited from program grants awarded by specialized organizations. The research was described in regular progress reports. Everything was done according to the usual scientific procedures. From Galton to Mengele, there was no discontinuity, no break. There is an imperceptible progression from a well-intentioned scientist, theorizing in his laboratory, to the criminal injecting formalin into the hearts of Jewish twins or gypsies so he could remove their multicolored eyes, or inoculating children with typhoid so he could compare reactions in mono- or dizygotic twins.

In this era of genetic engineering, the Human Genome Project, embryonic research, and sociobiology, we cannot forget. It is not possible to act as if nothing happened in the camps of Nazi Germany. What matters here is not the role of the physician who performed what he called "experiments" in the camps. It is that of the scientist

8. Benno Müller-Hill, *Murderous Science: Elimination by Scientific Selection of Jews, Gypsies, and Others, Germany 1933–1945* (Oxford: Oxford University Press, 1988).

who inspired the theory. It is the responsibility of those who advanced the doctrine on which the crudest possible version of biological determinism was founded. With the wisdom of hindsight, it is easy today to recognize that most of the ideas that inspired the eugenics movement were unjustified. And yet many of its followers were perfectly respectable men of science who thought they were acting in the public interest. So where did they go wrong?

Where they went wrong was in not examining critically enough the very concept of eugenics and what it implied. In particular, they did not correctly evaluate its social consequences. The danger for the scientist is to not test the limits of his science, and thus of his knowledge. It's to mix what he believes and what he knows. And especially, it's the certainty of being right. Geneticists did not adequately expose their ideas about eugenics to nonscientists. They were not sufficiently in contact with the rest of society before proposing a doctrine whose application would profoundly affect society. Moreover, scientists often work with abstractions, with concepts. To proceed with his analysis, the biologist must often dismantle the organism he wants to study. He is interested in an "object," a "system"—organ, tissue, cell type, protein, gene, and so on. An object has no dignity. It has no rights. You can do what you want with it without asking its leave. Working with human beings is not at all the same. No experiment of any kind should be tried on human beings without their consent. The respect for and dignity of the human person must be preserved in all circumstances. Even when he is himself the object, the human being must remain the subject.

It has occasionally been suggested that we pursue only "good" research, research that is supposed to bring only benefits to the human species, and that we abandon "bad" research, that is, research that might cause problems. You'd have to misunderstand what science is to make such a suggestion. Research is a never-ending process whose evolution we cannot predict. Unpredictability is in the very nature of the scientific enterprise. If what we are going to find is truly new, then by definition it is something we cannot know in advance. There is no way to predict where a given area of research will lead and, consequently, what its potential applications will be. That is why we cannot simply choose certain aspects of research and reject others.

Another suggestion is sometimes made: stop genetic research. It will open doors that shouldn't be opened. Doors behind which may lie findings that could, for example, risk increasing racial tensions. This reminds us of Adam or Prometheus. But it isn't knowledge that's dangerous, it's ignorance. And it is hard to imagine, given all the cultural, political, and religious —not to mention scientific—differences, what power, in virtue of what arguments, would be in a position to close all the genetics laboratories in the world. Moreover, it would mean cutting ourselves off not only from "bad" genetics but also from "good" genetics. And in the future our medicine will be largely based on genetics. After the Second World War, for purely ideological reasons and totally disregarding thirty years' worth of scientific data acquired from all over the world, the Soviets decided that genetics was a bourgeois science to be banished from Communist countries. On Stalin's order, genetics was replaced by the lunatic theories of Lysenko. The results are well known. For several decades, the development of biology and its applications, both in agriculture and in medicine, was totally blocked in East European countries. They are still recovering from it.

So at what level do we separate the good and the bad in applications? In gene therapy, for example, which is going to flourish in coming years? "The only means of fighting a plague," says Albert Camus, "is common decency."[9] And decency is what should protect scientists from deleterious use of their science. Decency, here, means that sci-

9. Albert Camus, *The Plague,* trans. Stuart Gilbert (New York: Knopf, 1948), p. 150.

entists owe it to themselves to tell the truth. But it must be the whole truth and nothing but the truth. First and foremost, they must speak out and make themselves understood by the public. They must explain to their contemporaries what they are doing, what stage their science is at, what is new, what can be expected, how this slightly disquieting idea of gene therapy, which aims to treat genetic diseases by providing patients with working versions of their defective genes, gradually took shape. Geneticists must demonstrate the difficulties and hopes such a technique brings with it. They must point out two very different situations. In the first situation, cells from damaged tissue are removed—blood cells, for example—and a healthy copy of the affected gene is inserted before reinjecting the patient's own treated cells back into him. This treatment of single somatic cells is indistinguishable in principle from other well-established medical practices: prosthesis, graft, or organ transplant.

But a gene can also be injected in such a way that it becomes incorporated in all the cells of the body, including the germ or reproductive cells which a person then transmits to his descendants. Here the situation is much more complex. To obtain this result, you would in effect have to inject DNA into fertilized eggs before any division takes place. In preparing such embryos, necessarily via fertilization in vitro, you would need to obtain a series of embryos among which there would always be normal embryos, unharmed by the lesion. It would then be much simpler to choose these healthy embryos than to inject unhealthy ones. In other words, there seems little justification for gene therapy in this context. On the other hand, it is possible, using this technique, to add a new genetic trait, for example, a gene that could confer certain advantages on humans. This is currently being done in animals and plants. But in the case of humans, the aim is different. It violates the genetic heritage of humankind. The point is no longer to heal someone, but to modify him, to mold him. And all biologists seem to agree: avoid this at all costs. In any event, on no account is it for scientists to decide questions of this magnitude. It is a matter for society. For citizens. The role of the scientist is to present the situation to them, detailing the possibilities, advantages, and dangers. All that can be explained simply. And only public understanding of the problems tied to genetics can dissipate public fear of the unknown. The ideal situation would be for the public to become capable of imagining and discussing these issues as they have, for example, with abortion, euthanasia, or the use of extraordinary measures to sustain life.

Telling the truth isn't enough. The *whole* truth needs to be told. Nothing kept secret. This is where the responsibility of the scientist is greatest. He must not let anything of what he suspects about the possible applications or threats fall into the shadows. In the case of gene therapy, he must describe all the difficulties of the undertaking and the potential dangers. Talk about vectors. Point out the role of recombinant viruses. Demonstrate the possibility that injected DNA nestles in close to an oncogene, which could unleash a cancer. He must measure the possible risk and explain why, faced with the seriousness of the disease and for lack of any other therapy, this risk is worth incurring.

Finally, the scientist must tell nothing but the truth. It is a mistake if, in the effort to obtain grants or to shine, he goes on to promise the moon. To make people believe that tomorrow all diseases will be cured. It was via a trial intervention against certain cancers that gene therapy had its beginnings in the United States. People must not be made to believe that gene therapy will rapidly permit us to control all malignant tumors, or that a genetic disease is vanquished the minute the DNA sequence that causes it has been isolated. It isn't important for the public to know whether a given scientific theory—for example, the bacterial origin of mitochondria—is true or not. On the other hand, it is important to know if one could, via gene therapy, cure a patient of Duchenne's myopathy or of cystic fibrosis. Similarly, it is important for the public to know whether the mental capabilities of individuals behave like simple Mendelian traits; that is, whether they are biologically determined. When he moves into a field of immense social consequences, the scientist has a duty to be particularly careful about what he says. It is important to clearly show the limits of genetic analysis without extrapolating it to domains for which information is still incomplete.

We will forever remain the victims of Zeus and Pandora. In locking all the evils of the world into the box that Pandora opened, Zeus forced humankind to fight to survive—to show imagination and ingenuity to protect itself from cold, from hunger, from illness, and various other perils. He condemned human beings to never-ending research.

7   Beauty and Truth

....

"If you want to find out anything from the theoretical physicists about the methods they use," said Albert Einstein, "don't listen to their words, fix your attention on their deeds."[1] Most people regard scientific research as a purely logical process, a cold and rigorous activity, as cold and rigorous as it appears in science textbooks or in books about history and philosophy of science. For their part, the philosophers discuss the hypothetico-deductive method ad infinitum. They analyze the process of discovery in detail. They talk about truth and "verisimilitude." Meanwhile, the scientists describe their own activity as a well-ordered series of ideas and experiments linked in strict logical sequence. In scientific articles, reason proceeds along a high road that leads from darkness to light with not the slightest error, not a hint of a bad decision, no confusion, nothing but perfect reasoning. Flawless.

1. Albert Einstein, "On the Method of Theoretical Physics," in *The World as I See It* (New York: Covici, Friede, 1934), p. 30.

And yet when you look more closely at "what scientists do," you might be surprised to find that research actually comprises both the so-called day science and night science. Day science calls into play arguments that mesh like gears, results that have the force of certainty. Its formal arrangement is as admirable as that of a painting by da Vinci or a Bach fugue. You can walk about in it as in a French garden. Conscious of its progress, proud of its past, sure of its future, day science advances in light and glory.

By contrast, night science wanders blind. It hesitates, stumbles, recoils, sweats, wakes with a start. Doubting everything, it is forever trying to find itself, question itself, pull itself back together. Night science is a sort of workshop of the possible where what will become the building material of science is worked out. Where hypotheses remain in the form of vague presentiments and woolly impressions. Where phenomena are still no more than solitary events with no link between them. Where the design of experiments has barely taken shape. Where thought makes its way along meandering paths and twisting lanes, most often leading nowhere. At the mercy of chance, the mind thrashes around in a labyrinth, deluged with signals, in quest of a sign, a nod, an unexpected connection. It circles like a prisoner in his cell, looking for a way out, a glimmer. It vacillates endlessly between hope and disappointment, between exaltation and melancholy. There is no way to predict whether night science will ever become day science; whether the prisoner will emerge from the darkness. When that does happen, it's pure coincidence—a freak. It occurs without warning, like spontaneous generation, and anywhere, anytime, like lightning. What guides the mind, then, is not logic. but instinct, intuition. The need to understand. A passion for life. In the interminable interior dialogue, amid the countless suppositions, comparisons, combinations, and associations that work in the mind nonstop, a flame sometimes rips the darkness, suddenly illuminating the landscape with a blinding light that is terrifying, stronger than a thousand suns. After the first shock comes an exhausting struggle with the old ways of thinking. There is a conflict with the universe of concepts that direct our reasoning. Nothing yet entitles us to say whether the new hypothesis will develop beyond its initial, crude sketch and become refined, perfected; whether it will stand the test of logic; whether it will be accepted as day science.

When he sits down to write an article to publish the results of his work, the scientist, consciously or not, forgets night science and talks only about day science. He has to organize a mass of data gathered over the course of months and years. He has to give these data a form from which he can extract a reasonable story that will become the official account of the research. It must be a story strong enough and persuasive enough to convince his colleagues; to get them to adopt his point of view, and even shed light on their own research.

Strange exercise in truth. Science is above all a world of ideas in motion. To write an account of research is to immobilize these ideas; to freeze them: it's like describing a horse race from a snapshot. It also transforms the very nature of research; formalizes it. Writing substitutes a well-ordered train of concepts and experiments for a jumble of untidy efforts, of attempts born of a passion to understand. But also born of visions, dreams, unexpected connections, often childlike simplifications, and soundings directed at random, with no real idea of what they will turn up—in short, the disorder and agitation that animate a laboratory. Still, as the work progresses, it is tempting to try to sort out which parts are due to luck and which to inspiration. But for a piece of work to be accepted, for a new way of thinking to be adopted, you have to purge the research of any emotional or irrational dross. Remove from it any whiff of the personal, any human odor. Embark on the high road that leads from stuttering youth to blooming maturity. Replace the real order of events and discoveries by what would have been the logical order, the order that should have been followed had the conclusion been known from the start. There is something of a ritual in the presentation of scientific results. A little like writing the history of a war based only on official staff reports.

At the end of the Renaissance, the sciences assumed their modern form. At that time, Western man was changing his relationship to the world profoundly, endeavoring to better use the evidence of his senses to construct the universe that surrounded him. It was during the Renaissance that Western art took a direction radically different from that in other cultures. In a few generations, with the invention of light and perspective, of expression and depth, Europe modified the very function of painting. Up to that point, painting had symbolized. From now on, it would represent.

When you visit a museum, you see a series of successive efforts that recall those of science. From the primitive to the baroque, painters have never ceased to improve their ways of representing, of showing creatures and things as faithfully as possible. Using optical illusion, they elaborated an entirely new world, a world open on all sides. There is a real break between a Madonna by Cimabue, immobilized in her veils in front of a symbolic landscape, and a woman by Titian, stretched out free and naked on her bed. This break corresponds to the one that separated the closed world of the Middle Ages and the infinite universe described by Giordano Bruno. Actually, the change observed in painting reflected the general upheaval brought about by the political conquest of the globe and the new view of the world that Western man had begun to construct. Between the thirteenth century and the classical age, Europe transformed its entire culture. Not only was symbolism replaced by representation, but monody by polyphony, prayer by action, mystery by drama, the narrative by the novel, the chronicle by history, and myth by scientific theory. Nor should we forget that the structure of the Judeo-Christian myth was largely responsible for the development of modern science. The former was founded on the doctrine of a prevailing order in a universe created by a God who was himself not part of nature, but who directed it by means of laws intelligible to human reason.

In some sense, science and myth play similar roles. They both answer a need of the human mind, supplying a representation of the world and the forces that control it. To avoid unleashing anxiety and schizophrenia, this representation must be unified and coherent. And when it comes to unity and coherence, science simply cannot compare with myth. In fact, science seems to be less ambitious. It doesn't try to explain everything straight off. It limits itself to clearly defined questions. It addresses well-circumscribed phenomena which it tries to explain with the help of detailed experimentation. It knows today that its answers can only be partial and provisional.

In contrast to science, other explanatory systems—magic, myth, religion—endeavor to be universal. They have answers to all questions, in all domains. They unhesitatingly describe not only the present condition of the universe but also its origin and even its future. Granted, many people do not accept the kinds of explanations that magic or myth supply. But who could deny their coherence and unity, considering that they require only a single, a priori argument to answer all questions and to solve all problems? Although very different, all explanatory systems, magic as well as myth or science, operate according to the same principle. The point, Jean Perrin[2] has said, is always to explain the visible world as a product of invisible forces; to account for what we observe by what we can imagine. Thunder can be viewed as a sign of Zeus's anger or a difference in electric potential between ground and clouds. A disease can be seen as the result of bad karma or an infection by a germ or virus. But in all cases, the phenomenon observed appears as the visible effect of a hidden cause related to the invisible network of forces believed to control the world.

2. Jean Perrin, *Atoms,* trans. Dalziel Llewellyn Hammick (London: Constable, 1916).

As I have already said, science seems, at first glance, to be less bold than myths, as much for its questions as for its answers. Actually, modern science is considered to have begun when its practitioners stopped asking, where does the universe come from? What is matter made of? What is life? and began to ask, how does a stone fall? How does water flow in a tube? How does blood circulate in the body? The change was amazing. Asking general questions had always led only to limited answers. By contrast, asking limited questions turned out to provide more and more general answers.

Whether scientific or mythic, the view of the world that man constructs is always largely a product of the imagination. It is popularly believed that to do scientific work, one need only observe and accumulate experimental results until a theory emerges from them. Nothing of the sort. One can contemplate an object from every angle for years, and never produce any observation of the least scientific interest. Arriving at an observation of value means having some idea about what there is to observe at the outset. Scientific advances often take place when an unknown aspect of things abruptly comes to light; not necessarily because of the advent of a new instrument, but owing to an original way of considering things, of looking at them from an unexpected angle, with a new perspective. A look that is always guided by a certain conception of what "reality" is or could be. No useful observation emerges without some idea of the unknown, of the region that lies beyond what experiment and reasoning entitle us to believe. In the words of Peter Medawar, scientific inquiry always begins by the "invention of a possible world, or a tiny fraction of that world."[3]

Mythic thought begins the same way. But it stops there. It constructs what it considers not only to be the best of worlds but also the

3. Peter Brian Medawar, "Science and Literature," in *The Hope of Progress* (London: Methuen, 1972).

only possible one. After which it easily fits reality into the framework it has created. Every event thus becomes a sign produced by the forces that control the world and, by the same token, demonstrate their existence and their role. With science, however, imagination functions only at the beginning of the process. Thereafter science must examine itself; expose itself to experimentation, to criticism, to refutation—in short, limit the part of dreaming in the representation of the world it constructs. Science is capable of imagining many possible worlds. The only one that interests it is the one that exists and that has proven to work for a long time. To keep the imagination from running wild, science ceaselessly attempts to reconcile the possible with the actual.

In the Western world, science and the arts have almost always flourished together in time and space. Sometimes astonishing convergences arise between certain aspects of the arts and sciences, for example, between literature and the study of the living world, from the end of the Renaissance to the Romantic revolution. What we call the "classical age"—the seventeenth century—is known above all as the era of representation, that is, of the study of forms and their organization, in all domains. This applies to visible forms, and obviously to painting, where we find the very symbol of representation in Velázquez's "Las Meninas," "the representation . . . of Classical representation," in the words of Michel Foucault.[4] Certain representations of religious scenes might even move the viewer to prayer and worship, which demonstrated a belief in the reality and efficacy of the images as if they were true relics. Similarly with musical or verbal forms, once the tools of language were defined and the structure of discourse analyzed. During this period, one of the principal forms of literary expression was the theater, presenting comedy and tragedy. In their works, Shakespeare, Molière, Racine, and Calderón de la

4. Michel Foucault, *The Order of Things: An Archaeology of the Human Sciences* (New York: Vintage Books, 1994), p. 16.

Barca created a range of characters whose behavior—what the audience saw and heard—revealed their personality. At the same time, natural history was endeavoring to classify plants and animals according to their visible structures, that is, based on the evaluation of what their surfaces revealed. In both cases, the analysis was founded on the external envelope of things, on what was perceptible from outside. The end of the eighteenth century brought about a total change in perspective. In both domains, the focus of interest became internalized, so to speak. On the one hand, poets and novelists who represented the literary mainstream began to speak about themselves, to describe their interior selves. They shifted the center of gravity. They expressed the state of their souls. On the other hand, naturalists began to be interested in the common properties of all living things. Below the visible surface of animals, they remarked the presence of an "organization" that governed the relations between the parts and obliged the organs to cooperate in coordinating vital functions. They spoke of life and sought to define what separated it from death. It is remarkable that the word "biology" appeared simultaneously in several places, precisely at the moment of the first literary suicide: that of Young Werther.

At the end of the nineteenth century, another change occurred that modified the most diverse aspects of Western thought and challenged the reigning order. Even the "me" became vulnerable when Freudian psychoanalysis exploded the self, smashing it to smithereens, as it were. The new understanding made it impossible to have any simple and direct recourse to reasoning and logic in explaining a fragmented personality that appeared more and more mysterious to itself. In literature, the very status of language was called into question, especially when the poet Stéphane Mallarmé spoke of a dissociation between the word and the thing to which it referred. The word "rose," for example, had neither color, nor odor, nor thorns. It did not bloom. You could not cut it. It was only a symbol that had nothing to do with the flower it represented. And it was precisely this void, this "absence of the thing," as Mallarmé put it, that gave language its suppleness and strength. It was because the word was entirely dissociated from what it referred to that it was free to interact with all other words, to combine itself with them in an infinity of sentences that constitute language.

The awareness of a void at the very heart of language caused a crack in the meaning of this language, in its relationship to the fullness of the world. The result was a kind of disequilibrium or destabilization, which led to profound changes in the arts of the twentieth century. In literature, efforts to get rid of conventional writing and its customs would lead to the absurd of the existentialists and the nihilism of Beckett. In painting, the disappearance of classical representation and the explosion of color pointed the way to surrealism and abstraction. In music, a deemphasis of melody, repetition, and meter would ultimately abolish length and anticipation.

In the sciences, the end of the nineteenth century would prove no less destructive of the old ways of thinking. I do not intend to describe here what everyone knows: the difficulties caused by the rigid determinism that prevailed since Newton and Laplace; the impossibility of predicting certain events, like the behavior of a gas molecule or the genotype of a future person, two situations that depend purely on the statistical analysis of large populations; the role of history and, consequently, of chance in the evolution of complex objects, especially in the living world, but also in the nonliving world. The twentieth century brought new limits to our ability to formalize and to predict, which is illustrated by a whole collection of words: instability, chaos, relativity, quanta, undecidability, indeterminacy resulting from the effect of the observer on the phenomenon being observed. The break between strict determinism and the blind process of evolution in the sciences thus corresponds with the collapse of meaning, of forms, and of intelligibility in the arts.

These examples suggest that, at certain moments in a history and culture, a kind of echo resonates between the way artists and scientists orient their thinking and the images they use. As if some force inclined everyone's efforts in the same direction. And yet these kinds of convergences are quite difficult to analyze. What is possible in different domains is often limited by the range of current opinion and the network of beliefs, knowledge, and attitudes that characterize a culture at a given time. This network is never arranged in a totally logical way and never constitutes an entirely coherent system. Indeed, the beliefs that exist at any moment are rarely in total agreement. They are usually independent, if not contradictory. How, for example, do we logically reconcile the doctrine of free will with that of destiny, or of a direction to history? Or even the idea that a work of art can express the most intimate, most personal part of a person and still have a universal character? Beliefs that are rooted most deeply in a culture very often have no logical basis. They were not adopted consciously. The same applies to the most fundamental concepts—notions of time, space, and causality—that guide our perceptions and shape the view that we have of the world and of ourselves.

Whether social or intellectual, whether it arises in politics, in art, or in science, every revolution demands above all a shifting of possibilities, a rearrangement of the system of beliefs. But the origin of these changes is more often than not difficult to pinpoint. Relationships between ideas and cultures, between beliefs and practices, constitute not unidirectional processes, but complex sets of interactions—it is the perpetual question of the chicken and the egg.

What a scientist does is thus determined not only by his own view of the world but also by that of his times. A sampling of scientific writings from a certain period reveals, strikingly, that everyone is talking about the same things and says more or less the same things about them, even those who are in total disagreement with each other. The same is true for artists. To be persuaded, one has only to listen to various pieces of baroque music, which all have a family resemblance. Or to visit a museum and see that in the sixteenth century all the Dutch painters were painting the same landscape, and that in the nineteenth century, all the English painters were painting the same portrait of the same women.

"Only the truth is beautiful," affirmed Nicolas Boileau in one of his *Epîtres*. "Only beauty is truthful," confirmed Anatole France in *La Vie littéraire* (The Literary Life). John Keats, in the "Ode on a Grecian Urn," went them one better: "Beauty is truth, truth beauty." Although there is beauty in these words, their truth may be less obvious. The sciences seek to construct a view of the world as close as possible to what we call reality. It is a collective enterprise in time and space. The arts aim to produce representations of the world, each of which expresses a personal view of reality such as it is perceived or imagined or dreamed. Most of the time this enterprise is an individual one. What is true, however, is that beauty and truth vary across cultures and, within a culture, with time. The relationship between truth and beauty or, more generally, between science and art, is an old issue, always difficult to tackle. There are obvious differences, which have been much debated. They center on two principal themes.

1. Scientific work is inexorably linked to the idea of progress, the like of which is not found in art. A truly "perfect" work of art will never be surpassed; it is ageless. In science, however, everyone knows that his work will be overtaken more or less rapidly, because all scientific work generates new questions, which is in fact its function. In other words, Beethoven did not surpass Bach, nor Picasso Rembrandt, in the way that Einstein surpassed Newton. In the words of Victor Hugo: "Pascal the savant is outrun; Pascal the writer is not."[5]

5. Victor Hugo, *William Shakespeare*, trans. by A. Baillot (Boston: Estes and Lauriat, [19–?]), p. 96.

Yet, as Gunther Stent[6] emphasizes, correctly, here what is being compared is incommensurable: on the one hand, a work of art, on the other, the *content* of scientific work. A painting or a novel is a work of art. A scientific theory, in contrast, is not a work of science, but the content of a work, such as a book, an article, a conference, and so on. In a novel, for instance, agreement of theme and form, of content and style are what give the work its value. It is impossible to separate one from the other. The importance of the content can even—in poetry, for example—diminish to the point where the aesthetic character of the piece ends up being exclusively in the rhythm, in the music of the words. In science, on the other hand, it's pretty much exclusively the content that gives a work its value. And the content of an article or a scientific book can often be summed up in a few sentences.

2. A second difference often pointed out between art and science is that the scientist describes the exterior world, in which objects and events have an existence independent of the human mind. The objects and the laws are already there. The role of the scientist is limited to discovering them; to gathering them like apples from a tree; to presenting them, like a statue on the day of its unveiling. The artist, on the other hand, describes an interior world where objects and events have no reality, but appear like pure constructions of the human mind. The role of the artist is thus to create new objects that spring entirely from his mind, like Athena emerging fully formed from the forehead of Zeus. *Othello* is a creation; the structure of the atom a discovery. Hence a difference in the role of the individual. The author of a work of art is unique, irreplaceable. The author of a discovery is interchangeable. Without Gustave Flaubert, no *Madame Bovary*. Without Mozart, no *Magic Flute*. By contrast, if such-and-such a discovery hadn't been made by Professor A, it would have been made by Doctor

6. Gunther Stent, "Prematurity and Uniqueness in Scientific Discovery," *Scientific American* 227 (1972), pp. 84–93.

B, even by Mr. C or Mr. D. Without Newton, another physicist would have discovered gravitation. Without Darwin, Wallace would have proposed the theory of evolution. Most scientists agree with this point of view. They almost never use the words "creation" or "creativity" to describe their activity. They themselves think that their business is, above all, facts; that they bring phenomena to light; that they reveal external objects. Indeed, laypeople think that science does little more than record facts, the way a camera takes photographs of its surroundings.

But that's not how our brain works. The brain must have evolved as a function of very varied and very complex factors, including the ability to interpret the external world. Living beings are traversed by a triple flux of matter, energy, and information. Only in this way are they able to live, grow, and multiply. An organism thus has an absolute need to perceive its environment or, at least, the aspects of that environment that are connected to the organism's basic requirements. With evolution, perception sharpened and organisms were able to take in ever more complex information. Every organism, it follows, possesses sensory equipment that permits it to perceive certain aspects of the exterior world. Each species moves, so to speak, in a particular sensory world from which other species are excluded in part or in toto. The exterior world, as each species perceives it, is a function of its sensory organs in much the same way sensory and motor events are integrated by the brain. An organism never detects but a part of its environment, and that part is particular to that organism. All of this is also true for us. We ourselves remain locked in the view of the world that our nervous and sensory equipment imposes on us. To the point where we cannot imagine the world in a different way. We don't have the means, for example, to conceive, or even to imagine, the world in which a fly lives, or an earthworm, or a seagull.

Mammalian visual and auditory perceptions are integrated with the help of a spatial and temporal code that permits us to connect the origin of sound or light stimulus to common sources that persist in time and space. During waking, a formidable quantity of information is shipped via the senses to the brain of a higher mammal. If this quantity of information can be handled by the brain, it's because it is organized into masses, into entities that form the "objects" of the spatio-temporal world of the animal. It is also because the identification of an object is preserved even though perception repeatedly alters in space and in time. These entities, these objects are, for the animal, the elements of its daily experience.

Although the human brain may be the most complex, it is clear that it doesn't function simply by recording nature. If our senses had to furnish us with a complete image of the external world, we would all be completely overwhelmed. The brain seeks regularity in nature. The signals transmitted to us by our senses are organized in a way that gives them structure. The eye, for example, is not a machine for communicating to the brain exactly what it sees. Over the course of the last twenty or thirty years, neurobiologists have shown that it is wired to pick out boundaries, contrasts of light, differences in color, and so on. With every exchange between the eye and the brain, incoming signals are selected and ordered by the nervous system. Each step thus implies the selective destruction of information. It is this process of integration that predisposes us to identify certain types of regularities and leads us to find laws in nature that permit us to situate ourselves within it. These sensory and cerebral transformations are sufficiently alike from one human being to the next that we all see external objects in a similar way. But there are enough individual variations to allow us each to create a personal impression. To take it further, just as an artist chooses from among his observations, impressions, and memories what he judges useful for the work he is producing, so the scientist selects a subset of his observations and, from among the phenomena related to it, identifies those that seem pertinent to him. We might say that, for a given object, there are a multiplicity of possible descriptions, and for a given description, there are a multiplicity of possible presentations.

Clearly, then, a physicist's description of the atom is not an exact and fixed reflection of reality laid bare. It is a model, an abstraction, the result of centuries of efforts by physicists who devoted themselves to a little group of phenomena in an attempt to construct a coherent view of the world. The description of the atom seems to be as much a creation as a discovery.

As in literature or in painting, there is style in science. Not only a way of looking at the world, but also a way of questioning it. A way of acting with regard to nature and talking about it. Of concocting experiments, carrying them out, extracting conclusions from them, formulating theories. A way of shaping the experiments into a story, to tell or to write down.

Take Pasteur, for example. There was something exceptional in his style, something irresistible, swaggering. Something of the cavalry charge that led him to leap from one area to another. To go from chemistry to crystallography, then to the study of the least well known aspects of the living world. To fly without hesitation from the diseases of yeast to those of humans, confident of his strategy, his ability to deduce applications from theory—or the reverse, to squeeze the most abstruse theories from the most concrete problems. He had astonishing intuition; he generalized with wild audacity.

Molecular dissymmetry; fermentation; so-called spontaneous generation; studies on wine; diseases of the silkworm; studies on beer; virulent diseases; virus vaccines; rabies prophylaxis—Pasteur's works read like a series of victory bulletins. There was a military side to the man, a strategic side. There was something a little Napoleonic in his way of always taking the initiative, of abruptly changing course, now appearing where he wasn't expected, now concentrating his energies on a narrow area to the breakthrough, exploiting his successes, recognizing their importance, and even organizing his own publicity or forcing others to bend to his views. Like that of Napoleon, Pasteur's art consisted in always joining battle at a moment of his own choosing, at a place of his own choosing, on his own ground. And his ground was the laboratory; his weapons were experiments, protocols, culture flasks. Whatever new field he entered, whether he took up grapevines or silkworms, chicken cholera or rabies, Pasteur sought each time to transform the problem, to translate it into other terms, to make it accessible to experimentation. Today we do things no differently. All the activity of biologists tends to reformulate the most varied problems into questions amenable to the laboratory. All their efforts aim at posing questions which experimentation can answer. It is from Pasteur and his strategy that modern medicine began as well as what we now call public health.

Without Pasteur, someone else would certainly have discovered the role of germs in infectious disease. Someone else would have shown the existence of filterable agents, of what would later be called viruses. Someone else would have demonstrated the possibility of vaccines. But most likely under very different conditions. More piecemeal, over a longer time, involving many researchers from many countries. If this study had been done differently, not by one man and his team, in a single body of work—you might even say in one go—but here and there, by several laboratories, a step at a time, groping about at length; if the solutions had come in dribs and drabs, and not in a single élan, they would nonetheless have kept their fundamental place in the history of biology and medicine, but they would have appeared as only one important strand of work among others, work that fit the current mold of research; spectacular work, to be sure, but lacking the grandeur of the Pasteurian epic.

Similarly, without Einstein, there would still have been something like the theory of relativity; without Darwin, something close to the theory of evolution. But they wouldn't have been the same theories. They wouldn't have been formulated in the same way or presented with the same vigor, the same force of persuasion. They wouldn't have had the same influence or the same consequences. In science also, each work—not simply the content—is unique. But as in art, to restate George Orwell's aphorism, among all these unique works, some are more unique than others.

It is likely that the mental representation of the external world became richer over the course of development through the stages of encephalization that led to *Homo sapiens*. Once an integrated image of the spatio-temporal world had been obtained, in which one could see, hear, feel, and touch moving objects, once the permanence of these objects in time was assured, it became possible to store this representation. All these properties then made possible two of the most remarkable features of the brain. First, the brain is capable of breaking up stored images of past events into their constitutive elements; these elements can then be recombined in new ways to produce new images and new situations and scenarios. This capability allows us not only to preserve the images of past events but also to imagine possible events and thereby to invent the future. Second, by associating auditory perceptions of temporal sequences with certain modifications of the sensorimotor apparatus of the voice box, we gained the ability to symbolize and code cognitive representations in entirely new ways. According to this hypothesis, the primary function of language would have consisted in permitting a more detailed representation of a finer, richer reality. Just as many linguists maintain, the use of language as a system of communication between people would have occurred only secondarily.

The facility with which communication establishes itself between individuals is found throughout the animal world. Simple codes must have sufficed for communicating what was needed for sharing information about the essentials of life, even for hominids, who lived in groups and got together to hunt and to defend themselves. By contrast, the ability to recognize objects and events months or years later required a much more elaborate coding system, one that allowed translation of the representation of a visual and auditory world with sufficient precision and detail. The unique character of language comes not so much from the fact that it serves to communicate instructions to act, but that it permits symbolization and evocation of cognitive images. Man models his "reality" with the words and sentences he uses as much as with his sight and hearing. In the development of imagination, the plasticity of human language constitutes an unequaled tool. Owing to an infinite combinatorial system of symbols, it allows the mental invention of diverse worlds. Each of us thus lives in a "real" world that is elaborated by our brain with the information contributed by our senses and language. And this real world is the backdrop against which our daily existence unfolds. It is the theater of our life.

In art as well as in science, what is important is to try. On the one hand, to try contrasting colors or harmonic themes or combinations of words, then to reject what we don't like. On the other hand, to try things; to try ideas—every idea that comes into our heads, every possibility—one by one, systematically; then to throw out what doesn't work experimentally and accept what does work, even if it contradicts our tastes or our assumptions. Most of the time, these efforts don't lead anywhere. But once in a while a totally outrageous experiment opens a new avenue. For example, the bizarre idea that came to my friend Elie Wollman one day while we were trying to analyze bacterial conjugation: why not unceremoniously separate the happy couples by placing them in a kitchen blender—a sort of coitus interruptus for bacteria? The experiment had an unexpected result: it allowed us to show that, during the process, the "male" chromosome was transmitted to the "female" at a constant rate, like a string of spaghetti being swallowed by the female. Consequently, we were able to think about and analyze bacterial sexuality in a new way. The beginning of research is always a leap into the unknown. Validation of the initial hypothesis happens only after the fact. Wrongheaded ideas and eccentric theories abound in research. They are as countless as bad works of art. No one can say where research will lead.

"What is now proved was once only imagined," wrote William Blake in *The Marriage of Heaven and Hell*.[7] It is in the imaginative phase of the scientific process, as hypotheses are formed, that the scientist functions like an artist. Only afterwards, when critical evaluation and experimentation come into play, does science separate from art and follow a different path. A poem or a painting isn't like a scientific hypothesis. Nevertheless, the imagination is still the driving force, the creative element, in science as much as in art or in any other intellectual activity. It wasn't just a simple accumulation of facts that led Newton one day, in his mother's garden, to suddenly see the moon as a ball thrown far enough to fall exactly beyond the horizon all around the earth. Or that led Planck to compare the radiation of heat to a shower of quanta. Or William Harvey to see the beating of a mechanical pump in the isolated heart of a fish. In each case a previously incomprehensible analogy was suddenly grasped.

As Arthur Koestler[8] has pointed out, this way of thinking seems completely different from that of King Solomon comparing the breasts of his beloved Shulamite to a pair of fawns, or from that of William Shakespeare describing life as a "tale told by an idiot, full of sound and fury." And yet, despite the very different means of expression that characterize poet and scientist, the imagination operates in the same way. Often an idea born out of a new metaphor will be the one to guide the scientist. An object or an event is suddenly seen in an unusual and revealing light. This sudden illumination is some-

7. William Blake, *Blake: The Complete Poems,* ed. W. H. Stevenson. 2nd ed. (London: Longman, 1989), p. 109.

8. Arthur Koestler, *The Act of Creation* (London: Hutchinson, 1964).

times accompanied by a resounding "Eureka!" that simultaneously expresses intellectual lights going on and emotional shock. I will never forget Jacques Monod's burst of laughter one day in 1963. An enormous laugh you could hear over our whole floor at the Pasteur Institute. For several months, he had been concentrating on the properties of so-called allosteric proteins. That afternoon, it suddenly occurred to him that you could explain most of these properties if you accepted that the proteins were specific oligomers, that is, formed by an even number of symmetrically arranged subunits. He was playing with a big pair of dice, showing everyone in sight the virtues of these structures, which oscillated easily between two states, one with enzymatic activity, the other without. And when people asked him how he had come to his realization, he answered: "For a few weeks, I've been trying to think like an allosteric protein. And today, I suddenly realized, I felt throughout my entire body the enormous potential of such a symmetrical structure."

Imagination is the combination and manipulation in one's head of mental objects such as images, symbols, words, cognitive structures, and so on. In a variety of areas, the creative act often corresponds to an abrupt leap away from usual patterns of thinking to associate two objects whose combination up to that point would have been non-obvious. Rational, conscious thought isn't necessarily the best tool for mixing mental images or representations in this way. When the mind has been concentrating on a problem for a long time, calm and relaxation can sometimes be more conducive to stirring and mixing images and ideas, to combining apparently incompatible structures and discerning unsuspected analogies among them. Many scientists have described the experience of having suddenly found a solution they'd been seeking for ages under totally unexpected conditions: in bed, half asleep; on a bus; staring at the flames of a fire; playing with a child. I myself had an experience of this kind. One afternoon at the movies, with my wife, I was only half paying attention to a rather boring film when it suddenly hit me that the two kinds of work going on in our laboratory at the Pasteur Institute—the work on lysogeny with André Lwoff and that on enzyme-induced biosynthesis with Jacques Monod—were actually nothing but two separate aspects of the same phenomenon, two expressions of the same mechanism. And because of certain particularities of the phage system, in both cases the locus of regulation had to be the DNA itself. This realization struck me with shock and with absolute certainty. A certainty that at first was not shared by my colleagues. But of course for me, the insight expressed an obsession I had had for weeks, my attention fixed on the particular question, going round in circles, saturated by this problem. Right up to the moment when either chance or fancy associated two areas that we all had previously perceived as totally separate.

Little by little, step by step, the small child constructs its environment. Similarly, the scientist gradually constructs his reality. Science imitates nature no more than does art. It recreates it. In breaking down what he understands of reality to recompose it in another way, the painter, the poet, or the scientist each builds his vision of the universe. Each fashions his own model of reality in choosing to throw light on aspects of his experience he judges the most telling, and in brushing aside those that are uninteresting to him. We live in a world created by our brains, with continual comings and goings between the real and the imaginary. Perhaps the artist takes a little more of this and the scientist a little more of that. But it is simply a matter of proportion. Not of nature.

Conclusion

....

"Did science promise happiness? I don't think so. It promised truth, and the question is whether the truth will ever make us happy."

—EMILE ZOLA,
ADDRESS AT THE STUDENTS' ASSOCIATION
BANQUET, 18 MAY 1893

In a discussion between Confucius and one of his students, the student asked the Master what he thought was essential for governing well. The Master replied, "What is necessary is to rectify names." "So, indeed!" said Tse-lu. "You are wide of the mark! Why must there be such rectification?" The Master said, "How uncultivated you are, Yu! A superior man, in regard to what he does not know, shows a cautious reserve. If names be not correct, language is not in accordance with the truth of things. If language be not in accordance with the truth of things, affairs cannot be carried on to success. When affairs cannot be carried on to success, proprieties and music will not flourish. When proprieties and music do not flourish, punishments will not be properly awarded. When punishments are not properly awarded, the people do not know how to move hand or foot. Therefore a superior man considers it necessary that the names he uses may be spoken appropriately, and also that what he speaks may be carried out appropriately. What the superior man requires is just that in his words there may be nothing incorrect."[1]

Some words evoke fear. The word "eugenics," for example, troubles us because it connotes unacceptable behavior that led to the sterilization of individuals considered "inferior," before it was used to mask the horrors of the Nazi camps. Other words, like the word "race," have been robbed of their meaning and used as biological alibis for cultural excesses. Even the word "genetics" frightens many people, because it is used thoughtlessly to influence social policy. To maintain that intelligence is essentially inherited—in other words, controlled by the genes—is to say that social policies aimed at educating disadvantaged populations are pointless. Those who proclaim that boys have a gene "for" mathematics intend to say that girls do not. Bosnia and Rwanda taught us that genocide, which we thought no longer possible after the fall of Nazism, could perfectly well happen again. We also know that arguments demonstrating the existence of clear differences in aptitude among groups are often used to justify discrimination and to obstruct policies aimed at combating injustice. The presence of a genetic component in a human behavioral trait doesn't at all mean that this trait is uniquely determined by the genes. Today we know that the development of the human embryo brings into play an ongoing interaction between the genes and the environment.

We sometimes wonder whether there might be limits to scientific research. This question is relatively new. The eighteenth century

1. Robert O. Ballou, ed., "Analects of Confucius (Lun Yu)," in *The Bible of the World,* book 13 (New York: Viking, 1939), p. 411.

never even imagined the possibility of such limits. On the contrary, the prevailing view was that sooner or later science would resolve all the questions people raise. But clearly some questions have nothing to do with science. There is a limit to scientific investigation. Science declines to answer questions like, what is the meaning of life? How did everything begin? What is our purpose on earth? Faced with such questions, science has nothing to say. We cannot even imagine what type of scientific progress might make a response possible. An entire domain is totally excluded from all scientific inquiry—that which concerns the origin of the world, the meaning of the human condition, the "destiny" of human life. Not that these are idle questions. Each of us, sooner or later, asks them. But these questions, which Karl Popper[2] calls "ultimate," are a matter for religion, for metaphysics, even for poetry. Science cannot answer them.

If we confine ourselves to questions having to do with science, we might ask what sorts of factors could limit science. This question has been discussed by Peter Medawar,[3] who distinguishes two types of possible limitations. First, the growth of science might be halted by some property inherent in the very process of scientific research. For example, the research process might spontaneously undergo a progressive slowdown and stop of its own accord. It is conceivable that scientific development might be limited in the way the height of buildings is limited, in that they cannot continue to mount forever toward the sky. Or in the way the size of an animal, such as an elephant, is limited, in that it does not continue to grow endlessly in all directions. We might also wonder whether science is capable of exceeding a certain sum of knowledge. But, a priori, we don't see what could

2. Karl Raimund Popper, *The Logic of Scientific Discovery* (London: Hutchinson, 1968).

3. Peter Brian Medawar, *The Limits of Science* (New York: Harper and Row, 1984), p. 67.

limit knowledge in this way and force research to come to a standstill on its own.

A second possibility: there may be a limit to scientific knowledge reflecting human limitations. When we tackle a new domain, we first learn what is easiest. Only after that do we take on what is complex, what is difficult. This second stage requires more finesse, better instruments, sharper analysis. Two analogies are helpful in talking about our cognitive apparatus. When we go fishing with a net, the size of the fish that we catch depends on the mesh of the net. Our cognitive net could have a mesh too big to catch fish below a certain size. Similarly, the powers of the microscope are not due to its magnifying capacity, as in the case of a magnifying glass. What allows the microscope to reveal details is its resolving power. In the middle of the nineteenth century, the optical microscope was perfected to the point where it suggested various structures in the cell, but did not reveal their details. Viruses, especially, were not visible under the microscope and became visible only with the electron microscope. We might wonder whether there isn't some limit to the power of resolution of the human brain or the human sensory system. For the moment, we can't really imagine what might restrict our analytical power in this way. But you never know. The human brain might be incapable of understanding the human brain.

In addition to a possible limitation to what human beings *can* learn, we might also wonder about potential limitations to what we *should* learn. In other words, is there information that leads to knowledge which it would be preferable not to acquire? Is there, in scientific research, a limit imposed no longer by the capacity for knowing, but by the power conferred by knowing? Must we stop learning certain things out of fear of the use to which the knowledge might be put? It's an important point. For though we have often been urged to refrain from certain applications of science, it has seldom been claimed that knowledge itself should be avoided. When, at the end of the last century, Pasteur vaccinated sheep against anthrax, the farmers and the mayors of neighboring villages clamored that the fool had to be stopped before he destroyed all the local livestock. Fortunately, no one listened. When, at the end of the 1970s, ecologists tried to prohibit research on genetic engineering, their advice wasn't followed, and the whole of medicine today is based on research carried out since then. But in all these cases, we already had the knowledge; the argument centered only on its applications. Should we or should we not use genetically engineered plants, at the risk of transforming entire fields? Should we or should we not use bacteria to make useful proteins, such as growth factors or hormones, at the risk of producing monsters?

Underlying these questions is the larger doubt of whether we should even continue to acquire certain aspects of knowledge itself. For example, in human genetics, we can imagine that decoding the human genome could have potentially dangerous consequences. But that danger, too, if it exists, would in the end lie in the application of newly acquired knowledge, not in the knowledge itself. We cannot stop the quest for knowledge. It cannot be dissociated from the human species. For part of human nature is to seek to understand nature. As I have already said, we cannot predict in which direction research that is just beginning will go, nor what it might bring about. We cannot pursue what will become "good" science and stop what we might consider "bad" science. And if we cannot stop research, neither can we keep only a part of it. In any case, we have nothing to fear from the truth, whether it comes from genetics or elsewhere. What we have to fear is misrepresentation of findings and the distorted meaning that people give them.

More than three hundred years have passed since the birth of science in the West—since it first tried its hand in everything; since it was used to construct what we call modern civilization. Science became the source of all the elements of contemporary technology: those we like, such as airplanes, television, penicillin, birth control, and those we detest, such as thermonuclear bombs, pesticides, and many kinds of pollution. Three hundred years is not a very long time. But it is long enough to try to evaluate the process, to decide whether this approach has served humankind or not. On this subject, there has been some disagreement. From the beginning, voices have sounded in opposition. Voices that became more strident now and again—at the beginning of the Industrial Revolution, for example, or with the advent of nuclear energy. Voices that cried out, "That's it! That's enough! Stop everything! Let's go back! Let's find another system, something less dangerous to the human species!"

Of course, scientists have a different perspective. For them, the scientific enterprise represents humankind's greatest success. Together with the arts, science is what really allows the human adventure to develop its potential to the fullest. But what science has accomplished up to now is only a beginning. Indeed, science wasn't really born three hundred years ago. It has only been developing systematically for about a century. Only fifty years ago did it find its rhythm, become a kind of institution spread throughout the world, irrespective of borders, nations, languages, or religions. But at the same time, the more science advances, the more we realize how much still needs to be done. Biology, for example, is in its infancy. It has only just begun to exist, bringing in its wake a nascent medicine. What is on the horizon, if basic research efforts are sustained, is not only the control of many diseases or agricultural improvements. With a better grasp of the fundamental processes of the living world, we can hope to learn more about ourselves. We desperately want to know who we are, where we come from, and what we are doing here. Granted, science cannot answer all questions, as I have said. It can, however, give some indications, exclude certain hypotheses. Engaging in the pursuit of science may help us make fewer mistakes. It's a sort of gamble. But our alternatives are not so many. Moreover, today we number five billion. Tomorrow, we will number six billion. The day after tomorrow, twenty. Terrible problems lie in store for humankind. There, too, the pursuit of science appears indispensable in the search for solutions.

The main discovery during this century of research and science has probably been the depth of our ignorance of nature. The more we learn, the more we realize the extent of our ignorance. That in itself is major news. News that would certainly have astonished our ancestors in the eighteenth and nineteenth centuries. For the first time, we can face up to our ignorance. For a long time, we claimed to understand how things worked. Or we simply made up stories to plug the gaps. Now that we have seriously begun to study nature, we have a better idea of the breadth of the questions, of the distance we must go to begin to answer them. The biggest danger for humankind is not the development of knowledge, but ignorance.

I began this book by trying to explain why our condition is inexorably tied to the unpredictable, to the impossibility of answering the question that interests us the most: what will happen tomorrow? Not to know what tomorrow will bring affects each of us differently. Some people would like to know whether they will find a job, whether they will win at the races, whether their lovers will still love them, whether they will still be alive. Personally, what matters most to me is not knowing what the world will be like in five hundred years. Or a hundred years. Or even twenty years.

We are a formidable mixture of nucleic acids and memory, of desire and proteins. The century that is ending has been preoccupied with nucleic acids and proteins. The next one will concentrate on memory and desire. Will it be able to answer the questions they pose?

Index